ETA AND ITS JACKUPS

Pioneering, Engineering, and Making it in America

PETER LOVIE

Front cover images

Top Left:

Norbe I, an ETA Robray 300 Class jackup (Robin, 1979)

Top Right:

Ednastar, the first ETA Robray 300 Class jackup (Hitachi, 1976)

Upper Left:

Cast steel jackup leg joints at CFEM's yard in France

Upper Center:

Ednarina, another ETA Robray 300 Class jackup (Hitachi 1977)

Upper Right:

Author's hard hat: a souvenir from Southeast Asia of the 1970s

Bottom Left:

Dyvi Gamma afloat, leaving CFEM's yard in 1977

Bottom Center:

World's biggest jackups in 1977: *Dyvi Beta* and *Dyvi Gamma* on delivery at
Dunkerque, France (ETA Europe Class design)

Bottom Right:

Dyvi Beta, jacked up and operating as *Petrobaltic Beta* in 2018

Disclaimer:
The information and images provided have been attributed as best as practically possible. Sources for content in tables are identified where known. Best efforts have been used to give a historical idea of what was done and sometimes corroborated with recollections of former colleagues.

ISBN-13: 978-1-945532-87-0

Library of Congress Control Number: 2018958407

Printed in the United States of America

Published, Edited, and Cover Design by:

Opportune Independent Publishing Co.

113 N. Live Oak Street

Houston, TX 77003

(832) 263-1700

www.opportunepublishing.com

Author: Peter Lovie

Peter M Lovie PE, LLC

P.O. Box 19733 Houston TX 77224

Peter.Lovie@ETAanditsJackups.com

An email to discuss your ideas or point out any errors or typos would be welcomed. If you enjoyed this book and felt others might like it, please consider writing a review on www.ETAanditsJackups. com, Amazon, or on the website you purchased your copy.

CONTENTS

LIST OF PICTURES

LIST OF DIAGRAMS

LIST OF TABLES

FOREWORD

A recent rereading of an excellent history of the evolution of offshore drilling units and the offshore drilling industry (M. A. Childers, Chapter 14, SPE Petroleum Engineering Handbook, Volume II, Drilling Engineering) was a compelling reminder of how many people and companies played a role in the development of a highly advanced, technically sophisticated, international industry. It tells the story of how the offshore drilling industry began in the U.S. and evolved to become a worldwide industry working in harsh and challenging conditions through advanced technology.

Much has been written and recognized with respect to the technical side of the industry's remarkable evolution but less about the people and human side of the many technical breakthroughs and developments and how they became practical reality. As part of the industry's evolution, the personal and individual experiences of the industry's people not only deserve to be told, but are important to remember and record to understand how these advances ever occurred.

This is especially valid today in our modern world of technology, innovation, fast-paced change, and risk considerations. Many of the lessons apply to the commercialization of new technology, the role of both failures and successes, possible outcomes if something had been done differently, the crucial aspect of the way people work together effectively, how strategy is best applied and used, different styles of technology leadership, what brings inspiration and talented people together, the role of confidence and belief, and the resilience of the human spirit.

Even in the early days of the offshore industry's development, there were highly talented engineers and innovators drawn to the U.S. from other countries, and these people played a key role. My first job in the oilfield was with Kerr McGee in Oklahoma City. In this position, I learned how they pioneered the first truly offshore well out of sight of land, drilled in 1947 in the Gulf of Mexico, and acquired the *Breton Rig 20*.

The rig designer, John Hayward, was born in Liverpool, England in 1890. He served in the British Army in World War I, came to America, and became a U.S. citizen in 1931. He served in the U.S. Navy in World War II, and the offshore drilling rig he designed used his patent. Many thousands more from around the world would be drawn to the offshore industry, and the US as the industry became international.

It is important that the human side of the industry's story be told along with the story of its highly technical development: the dreams, pioneering and selling of new ideas, and business hurdles faced and overcome— frequently by individuals and small companies rather than big industry names.

Peter Lovie is an example of an industry veteran. Like others from around the world, he established a respected place for himself as a leader in an industry he came to love. He found adventure, excitement, challenges, the chance to work with a wonderfully diverse mix of talented, hardworking people, and the opportunity to pursue his passion for dreaming up better solutions and doing things professionally and right.

It is also the story of the company that Peter led as president and co-founded, Engineering Technology Analysts, Inc. (ETA), which started from scratch in Houston in 1970. Its initial business was very modest: consulting and applying new computer methods to perform structural analyses. Then Peter identified future growth in the jackup fleet and a need for better designs. He published that vision in *Oil & Gas Journal* on January 10, 1972 and went to work to make it happen. In 1974, he and his band of "brainy millennials" as he now calls them, secured a contract to design two of the world's largest jackups, which would go to work in the North Sea. The pace was breathtaking, like Silicon Valley has been in the modern era!

ETA broke new ground with their cast steel leg joints to reduce local stresses and fatigue and with their new designs of longer jackup legs for deeper water and harsher sea conditions. Peter and his partner founded ETA with a belief in its ability to show the industry a new way in certain areas of jackup design, as well as a willingness to accept and take on the odds associated with the risk of a new startup venture.

 ETA's patents illustrate an unsaid side to pioneering: many new ideas were developed, but only a few became reality. Four patents were never used, but one was used in many jackups of ETA's designs delivered during ETA's existence and after. Although ETA went out of business in 1977, jackups of ETA's design continued to be delivered, twenty-two altogether by Peter's count. Many are still working today. The success of ETA in its short life inspired other design sources to enter the jackup design market, as this book's data shows.

Peter's compelling account of his offshore industry career starts in his native Scotland before he ventures to the U.S. and Texas, what he considers to be the land of opportunity. It is an honest account from his own perspective of the good and the bad, acknowledging both success and failure, satisfaction and despair, outstanding achievement and what could have been, and what went wrong. Peter conveys how a clash of personalities can get in the way and lead to failure. He emphasizes the crucial importance of values and integrity to success in any venture or field.

Peter acknowledges that in the long run, the willingness to learn, take on risk, and move forward prevailed. This approach ultimately resulted in ETA's significant achievements in its designs and in its contributions to the industry. In 2018, Peter's contribution was recognized by the Offshore Energy Center (OEC) through his formal induction into its Hall of Fame as an Industry Pioneer.

ETA was fortunate to catch the wave when it did. The offshore drilling industry has always been cyclical, and is accordingly associated with volatility. Like other industry participants, ETA faced good times in the 1970s. The industry's history from its inception records cycles and unexpected swings and shows just how large the vagaries have been.

Industry cyclicality was a key driver in development of our strategy at Atwood Oceanics. This recognition and acceptance of the reality of operating in a cyclical industry and the observation of a pattern of unpredicted surprises enabled our team to manage successfully in the black over a long period of time and achieve leading value creation.

From my own perspective, it was an honor to be part of an industry with its own special culture and to work with talented people of all imaginable backgrounds in many areas of the planet. This book by my old friend and colleague Peter Lovie shares some common experiences and has been a pleasure to read— and to recommend to others. Both of us grew up and had our initial educations in other countries— Peter in Scotland and myself in Australia— before each of us came to the U.S. on scholarships for further graduate education and subsequently found our separate ways to the offshore industry, where we met.

As a highly respected colleague and industry member, Peter clearly records how fortunate and indebted he feels for the opportunities that came his way during a long career in such a special and exciting industry. If there is one message he would want to leave to younger readers, I know it would be to have confidence and "have a go" in whatever you do. He would probably add: "Only in America!"

—John Irwin, President & CEO, Atwood Oceanics Inc., 1992-2009

INTRODUCTION

From 1970 to1975, a new company in Houston, Engineering Technology Analysts, Inc. (ETA), created a new generation of jackup designs. This innovation challenged the much larger and well-established Big Four jackup designer-builders by offering greater load-carrying capability at lower steel weights and employing a new, patented leg design that incorporated cast steel joints. These advances were commercial game changers, opening the potential for more shipyards to build jackups in what had been a specialty business.

ETA was formed in early 1970 by a 29-year-old immigrant and Houston resident three years after he arrived in the U.S. He started from scratch without capital but with a Master's degree, a new professional engineer (PE) license in Texas, and what he believed was an opportunity. That entrepreneurial engineer, the force behind the designs, was Peter Lovie.

ETA's first big break came in 1973 with an agreement to design the ETA Robray 300 Class jackups for 300-ft. water depths for Far East service—cutting-edge conditions at that time!

Four years after startup, the company had grown a team of "brainy "millennials" and secured a contract to design the world's largest jackups under the world's most demanding standards for service in the North Sea. Built in France to the Norwegian DNV Class, the *Dyvi Beta* and *Dyvi Gamma* jackups were delivered in 1976 and 1977 with legs a record 508 ft. long. They were the first of harsh environment jackups in the North Sea, setting a precedent that others followed. They became the subject of industry studies led by the Det Norske Veritas (DNV), which confirmed the validity of the design. (One still operates in 2018.)

This book examines the improbable startup and growth of ETA, despite the conservative offshore drilling world. We will explore the thinking behind the designs which were built, as well as concepts that were awarded patents but did not make it past drafting. At a time of great change and unprecedented growth in the offshore drilling industry, the construction of the ETA-designed jackups around the world was a demanding process. Since jackup accidents occurred relatively frequently in the early 1970s, the industry scrambled to do better in design and operation.

I served as ETA's president during its startup, growth and successes from 1970 to1975 and left in late 1975. Afterward, ETA encountered circumstances which led to its downfall (we will delve into these later), and ETA collapsed in 1977, missing out on the second big wave in jackup business. Despite ETA's demise, jackups employing its designs continued to be delivered during 1978-1982, and several remain in service. Chronicled here are the twenty-two jackups built from ETA designs and those surviving today. In addition, leading jackup builders copied the ETA leg chord design.

ETA was ahead of its time in many ways. ETA led the way in a shift in jackup design sources with market penetration that encouraged other designers to enter the market. During 1978-1982, the world's jackup fleet enjoyed its all-time building boom before it reeled under the bust during 1983-1986. The fleet was stagnant for the next 30 years. There was some growth in the 2010s. Then, from 2015-2017, offshore drillers experienced chronic oversupply during the petroleum industry's latest and possibly the most severe downturn.

In 2005, offshore drilling pioneer Tim Pease emailed me to ask what had happened with ETA and its jackups. Back in 1968-1969, Tim Pease was chief engineer at The Offshore Company (now Transocean), where I had my first jackup engineering job in the offshore drilling world. He was my boss back then, a half century ago.

Tim Pease emailed again in January 2017, joined by a longtime friend and industry authority, Dr. Malcolm Sharples. In 1974, three years out of University of Cambridge in UK with his doctorate in engineering, Dr. Malcolm Sharples was a consultant at Noble Denton in London, charged with vetting the ETA Europe Class jackup design to give underwriters confidence in insuring jackups of this new design for drilling operations and field and ocean tows.

Both men are supporters of the Oilfield Energy Center in Houston and its Ocean Star Museum in Galveston. Dr. Malcolm Sharples is one of the Center's founders, and Tim Pease has been recognized as an Industry Pioneer in its Hall of Fame. These two industry veterans wanted to get the real engineering story of the advances in pioneering jackups. They knew their stuff and grilled me for information!

Long emails responses to their questions morphed into a document that grew and grew as the discussion progressed for months.

An early draft in May 2017 was posted to www.ETAanditsJackups.com along with copies of articles and patents from the ETA days. It stirred encouraging feedback from others in the industry. Checking details often indistinct in memories of 40+ years ago,

I sought input from former colleagues and other industry insiders. Their perspectives taught me sides of the ETA story I never knew.

Personal conversations truly added substance and accuracy to the history I sought. I got reacquainted with former colleagues over long lunches, talking about old times. They prompted me toward research avenues I hadn't considered, and the narrative continued to grow as I created more and more diagrams, pictures, and tables.

With more research and conversations, I realized that there was much more to the story than the engineering advances that ETA pioneered. In order to uncover the full history, I had to bring to light themes of pioneering, engineering and making it in America.

In November of 2017, I stumbled across a file box in the garage, one of many my wife had been after me to clear out and dump in the trash! It contained records from the ETA days: prints of original 1975 design drawings of ETA jackups, including leg design details and cast steel joint details. Designed for deeper water and tougher conditions than ever before, these designs had been controversial subjects at the time of their conception. Now it was possible to include specifics on the pioneering features in *Dyvi Beta* and *Dyvi Gamma*, the biggest jackups of the day. These drawings were from when drawings were actually drawn by people with pencils, on tracing paper on drawing tables, well before the days of Autocad!

The saga has come together in seven parts. Early reviewers of my drafts believed it had to be published, and that encouragement prompted me to follow through.

Part I delves into ETA's startup, growth, and successes. It is a tale of coming to America, adapting, learning the oilfield lingo, and making it in the conservative offshore world of Texas in the 1970s. I recall these fledgling days as a young immigrant, who built and led a team comprised mostly of other young people. In 1966, it was not easy to come to America; even with a good education, an employer's vouching for me, and a stable home country, it took about six months to complete the immigration vetting process. Today it is another world, to say the least!

ETA's explosive growth during 1970-1975 did not end well. The 50-50 owners had conflicts, which led to my departure. Part II examines 1976-1977, the collapse of ETA, and the designs after its downfall.

Part III explains the pioneering design philosophies that ETA established and actually applied in jack-up construction. These philosophies drew upon contemporary industry guidelines and standards. In this section, designs using an ETA patent illustrate some design features of the Robray 300 Class jackups.

With their record long legs, the two ETA Europe Class jackups allowed drilling in deeper waters. In turn, they demanded that the hull be designed so that it was stable while the jackup that looked so top heavy was floating and being towed to location. In a conversation in 2017, Dr. Malcolm Sharples spoke of how, in 1974, he had been surprised that ETA jackups could move with a hundred more feet of leg than other jackups, due to the new leg design and the advances in stability analyses. In a 1973 presentation of a technical paper at the annual National Meeting of the Society of Naval Architects & Marine Engineers (SNAME), ETA advanced the industry standards on assessing the floating stability of jackups. The significance of that important technical paper on this arcane topic is also addressed in Part III.

Part IV contains jackup design concepts that ETA created and patented but which never went beyond the paper stage. These include a one-legged production jackup design—a first which inspired four other major companies worldwide to attempt to emulate it in the following four decades. Currently a one-legged jackup is being built on speculation in China and looking for work today.

Part V identifies all the jackups built from ETA designs and gives the history of what has become of them, long after my six-year stint at ETA. For reference, this section has pictures of completed jackups, pictures of cast steel leg joints during leg construction, and pertinent diagrams and drawings. The history includes a number of much smaller jackups, built for client requirements in which ETA served as jackup design consultant.

The world's fleet of jackups has long had a boom and bust history, even more turbulent than the overall petroleum industry. Part VI explores the revolution in design sources for jackups during these tumultuous times (some of these changes are not widely recognized and have only been brought to light through records and memories. ETA had led the way in challenging the Big Four jackup design sources in 1970-1975. Post-ETA, the world saw many new jackup design sources (some designer builders, others independent design firms) enter the market in the 1978-1986 wave of deliveries, encouraged by ETA's quick progress and pioneering.

No longer were the established Big Four based in the U.S. Gulf of Mexico as dominant as they used to be. Jackups could be designed in one place and built in many parts of the world. A wider choice of independent design firms emerged, inside and outside the U.S. There was a permanent shift away from U.S.-dominated jackup builders to a majority in the Far East. In fact, the traditional jackup builders of the U.S. Gulf of

Mexico (Bethlehem, LeTourneau-Vicksburg and Levingston) went out of business in the late 1980s due to the shift in building capabilities and the long market downturn.

Part VII on "Closing Thoughts" addresses the opportunities taken and missed by ETA. This section offers teachings from these experiences while highlighting ETA's contributions to the industry.

For historical interest, this section draws a comparison between the world's largest jackup in 1977 and that of forty years later: the forty-one-year-old *Dyvi Beta* and the three-year-old *Maersk Intrepid* respectively. Coincidentally, both jackups work in the North Sea today.

Part VII closes with personal reflections on the ETA adventure, how industry leaders influenced the thinking that went into ETA.

Some of the people who were at ETA are no longer around; they left the industry, retired, or passed away. In 2017 and 2018, I have been fortunate to have good discussions with several who were there when the ETA jackups were being designed and built, and they are in Acknowledgements.

Hitherto there has been no story done on ETA. Early on in my writing process, I learned there were at least three former ETA engineers who claimed to be the real brains behind the ETA designs: one in the Far East claimed to have designed the ETA Robray 300 Class design; another claimed on a website he was "the principal architect in the design and construction of the jackups *Dyvi Beta* and *Dyvi Gamma* for Norwegian drilling contractor Dyvi A/S; a third engineer claimed technical leadership behind ALL the designs and new technology (he died three years ago).

ETA must have been doing something right to generate all that fake news and résumé inflation!

The reality was that a number of very talented engineers worked together very hard at ETA, and no individual can claim complete responsibility. Rather, the credit is due to the team members and team leader.

In the early days, people in the conservative offshore world naturally wondered about the reliability of groundbreaking designs from a new company. It was understandable for the industry to hesitate to accept new designs from ETA as a small upstart design source. In 2017, the true depth of their concerns came to light as the American Bureau of Shipping (ABS), Det Norske Veritas (DNV), Noble Denton, and others imposed multiple and unusually rigorous "extreme vetting" of ETA's designs. That's right—"extreme vetting" existed in the offshore world long before Donald Trump used that term in 2017!

The long life of the ETA-design jackups has confirmed the design performance. Many jackups that used ETA designs still exist after more than forty-one years in operation, after most of their designers have retired!

Writing this book while nearing the end of my career, I recognized the importance of coming to America and the opportunity I pursued in founding ETA. In talking with industry colleagues, it turned out that the U.S. offshore industry has multiple leaders from other parts of the world for example, John Irwin (who wrote the foreword to this book) had great success as the CEO of Atwood Oceanics, Inc. after he came to America from Australia. Gordon Sterling, who retired a few years ago from senior management in Shell in the U.S. Gulf of Mexico, was also an immigrant - from Canada. At Shell, he pioneered multiple game-changing technical developments and also started here about the same time as I did, and reminded me that it was a time when engineers used slide rules and referred to their copies of Roark and Timoshenko!

Before me, two Peter Lovies attempted to make their way in America. In 1912, the first Peter Lovie, my paternal grandfather, who was a stonecutter in Aberdeen in Scotland, travelled to Vermont to build a new life, with the hopes of bringing over his wife and two small sons. Months later, during the Fourth of July holiday weekend, he died in a drowning accident. His wife and their two sons (the elder was my father) were not able to live out their father's dream and stayed in Aberdeen, Scotland.

My grandfather's brother Frederick finished his MA at University of Aberdeen, volunteered in 1915, and went to France in a regiment of the Gordon Highlanders. In 1918, he was awarded a Military Cross and then was invalided out before the Armistice. He completed his training to be a minister of the Church of Scotland and was selected by a church. He married, then had to resign from his job, apparently due to war injuries and what we now call PTSD. He died in 1924. These were not good years for the Lovies.

In 1928-1930, my father, the second Peter Lovie, was inspired by his father and his Uncle Frederick: attended University of Aberdeen (scholarship and working nights at The *Aberdeen Press and Journal* newspaper) and on to New York City on scholarship for graduate work at Union Seminary at Columbia University. Family difficulties back in Scotland meant he had to return after graduation, unable to follow his and his father's dream.

And so from that history I grew up hearing my father's anecdotes about New York, reading the international edition of *Time* every week. My parents, sister and I lived in Culross, Fife in the east side of Scotland, where many of the houses were built in the 1600s. We went to Dunfermline High School, founded in 1468. On Sunday evenings the family often listened to "Letter from America" by Alistair Cooke on BBC radio.

A few years later on one truly memorable summer afternoon in 1960, cheering Scots filled the Ibrox stadium in Glasgow to capacity. It was not to see the Rangers play football (i.e. soccer) but to listen to America's legendary Louis Armstrong! The newspapers had a field day. The *Scottish Daily Express* was full of pictures of the event, one showing Peter Lovie and his date in the enthusiastic crowd around where the musicians came out onto the field.

When the musical "West Side Story" came to play in Glasgow, I was still a student at University of Glasgow and took my parents. It was a big deal. In one song, the characters sang "I like to be in A-mer-i-ca," little did I know then that two years later I'd win a scholarship for graduate work at University of Virginia and I would be the Peter Lovie to ultimately enjoy the American Dream in Houston.

When the Oilfield Energy Center informed me that they planned to induct me into their Hall of Fame as an Industry Pioneer in 2018, it finally felt that this generations-long desire had come to fruition.

PART I

ETA STARTUP, GROWTH, AND SUCCESSES

The story of ETA and its jackups has its roots in Peter Lovie coming to America three years before the start of ETA and so the company's history is told here via his anecdotes.

As an engineer first (as well as president of the new upstart outfit that came to be composed of mostly what would be called "millennials" today), it was an adventure to challenge the jackup designers of the day. It might have been a micro version of what Silicon Valley pioneers would do generations later.

Today's research shows a total of twenty-two jackups based on ETA designs entered service in different parts of the world during 1976-1982: eleven for maximum depth exploratory drilling duty in 300-to-350-ft. water depths, one for 200 ft., and ten for lighter production drilling and workover service in shallower waters of 150 ft. and less.

I was very much in the middle of it all, laying out equipment and systems as well as thinking out new configurations. Then devising the right engineering models for analyzing and designing our jackups to carry more variables load, work in deeper waters, and tow stably, all with what we believed were lower structural weights than competitive designs, implying economies, in a quest for more performance for less steel. We believed we could do it all better than what was out there!

These achievements could only have been accomplished with the dedication and talents of the engineers and draftsmen who joined ETA as the company grew. I would coach them on what was needed in this rare new world of operating offshore with jackups and on what our designs were really looking to achieve.

Once these bright engineers got going, they really blossomed and excelled, building an exceptional body of expertise in the company. We did not have grey-haired experience but were determined to compete in the marketplace by employing our engineering brainpower and our drive for excellence.

More broadly from today's perspective, the two waves of jackup deliveries in 1970-1977 and in 1978-1986 were times of great change in the offshore drilling business; ETA was the new jackup designer at the front of the first wave, with many new shipyard–designers and new independent design firms entering the business during the even larger second wave of 1978-1986.

Coming To Texas And Learning The Lingo

In 1967, I emigrated from my native Scotland to Houston for a new career and life. The company I worked for in the U.K. was a division of the Stewarts & Lloyds steel group which had a long name as a supplier of oilfield tubulars. My employer was being shut down. Cameron Iron Works (now part of Schlumberger) had recruited me.

While waiting for my U.S. permanent resident visa and the vetting (all the forms, medicals, and interviews at the U.S. Embassy in London), I worked at Cameron's plant in Livingston, Scotland. There, I learned about a fellow Scotsman who had had applied as a chauffeur to pick up customer and company VIPs from the airport and look after them during visits to the plant. To his great surprise and delight, he was immediately hired on the spot because of his name: Sam Houston! Someone had realized how great it would be to have Texas based customers being greeted by someone who really was Sam Houston!

During that waiting period there was a terrific bang one November night when the frame in one of the forging presses cracked under load. The incident meant halting production for many millions of dollars of business. I was assigned to diagnose what caused the structural failure. It was simple—brittle fracture in the 3 in. thick plates of ASTM A212 plate in the forge press frame. No one had thought that the plates of the press frame would be exposed to cold winter temperatures as well as full loads during startup in Scotland. Whether in Houston or in a normally hot forge shop in Scotland, the expectation was that temperature would be high enough to avoid getting close to the transition temperature and hence that cracking. It was a dramatic experience that stuck with me–simple material properties, designs and temperatures could have big time operating effects. I learned later that that brittle fracture effect probably accounted for loss of several freighters in the North Atlantic in World War II.

On February 14, 1967 I arrived in Houston. About a month later I drove out Hempstead Highway to visit Cameron's new plant site. I was told it was "way out there in the middle of the bald-assed prairie." New to America and Texas, I had not expected a prairie in the middle of Texas and could only imagine what it would be like.

Coming to Texas was truly an experience. At the office block where I worked, a guy came to work in a Chevrolet Camino (not a pickup truck but the sedan with a truck bed back). He kept his deer rifle with a scope on hooks in the back window. Every day he wore boots and a western hat to work, just like a cowboy in the movies.

In Scotland, I had never lived on a farm or needed a firearm. I couldn't get a license for a 0.22 rifle or shotgun. It had only been big landowners and big shots that hunted. Yet here in Texas, an average working guy in a big city could own a high caliber deer rifle and hunt!

No longer did I zip around in a little Triumph Spitfire sports car with a four-cylinder engine just over a liter; instead, I floated around in an automatic drive Pontiac Grand Prix with a V8 engine. It had power steering and windows, six times the engine capacity, and the luxury of air conditioning!

In Houston, there were many big, new apartment complexes to choose from, all with a swimming pool. I settled in the Three Fountains Apartments, freshly built. It was a luxury unlike anything in Scotland, with pool parties every summer weekend. And that is how I discovered what people at work had really meant about the social life around the pool! Life was good!

There was certainly some adjusting that had to be done. The engineering office I worked in needed a new draftsman, and I was assigned to do the initial interviewing. One applicant said he came from Palestine, and I commented how that must have been a big change and long way to come to Houston to seek work. He looked at me kind of funny and said nothing. I was puzzled because he did not have a Middle Eastern accent. Later that day, I was bluntly told "That boy's no ay-rab, Palestine is an itty bitty town in the piney woods between here and Dallas." My Texas education continued.

The forge shop next to the office was really big with a high roof. Today, that forge plant is gone, the air is clear, the ground covered with a shopping and entertainment mall. But at the time, it was thunderously loud, smoky, and hot, and the earth sometimes shook with the thumping of the forges. A burly worker with a helmet, visor, and leather apron used a huge pair of tongs to guide hot, yellow chunks of exotic nickel cobalt alloys, which were suspended from an overhead crane, from one forge furnace or press to the next. The heat and pressure created discs of metal for critical service as jet engine turbine wheels. It was as if Dante's Inferno was on Silber Road. It was quite another world from the small farming and coal-mining town where I had grown up in Scotland.

On the job, I not only had to get used to the differences in culture and landscape, but also learn the lingo of the oilpatch. If you were a new hand who had never been on a drill floor before, you were a weevil. It was a derogatory term, but for good reason: your lack of experience could be a danger to yourself and to others. If someone was a weevil who didn't learn how to one day become a good hand, that unfortunate soul did not know thin shit from Shinola. It took a little questioning to understand what Shinola was: an old brand of boot polish.

Things have changed in the last fifty years. In 2017, Shinola has a website (www. shinola.com) that markets watches and fine goods from Detroit. It's said to be one of the coolest brands in America, a long shot from how we used that term in the oilpatch.

Within these new words, I also picked up an engineering design principle I'd never heard of, even in graduate school in the U.S.: the concept of building structures "hellferstout." Sometimes operational considerations on offshore drilling rigs were not amenable to rational structural design. However, they had to be robust enough to withstand wear and tear. Structural elements would therefore be built far stronger than an engineer might feel necessary, hellferstout to ensure that there was never any risk of structural failure in use.

Jackup drilling rigs also had unexpected critters on board, if only in name. Each rig has its mouse hole and its rat hole. Up on the derrick, there was a monkey board. More mysteriously, they also had a possum belly tank!

Growing up, I'd always understood that a spud was slang for potato. But it was different in the Texas oilfield, where jackups had spud tanks, and offshore drilling rigs would spud a well as they started operation.

There was an erotic-sounding process of nippling up… and the important matter of understanding leg penetration.

Despite the colorful local expressions, I was delighted to be a resident alien with a green card (it was actually blue and black). However, being a resident alien meant I had to register with the U.S. Selective Service a few months after arriving in Houston, I received a draft card but thankfully being just over 26 and with the country not then needing large numbers of conscripts for ongoing conflicts, I did not have to worry about being drafted for military service.

At work, I got a taste of the Southern work ethic. When people saw willingness to work, friendly encouragement abounded. Learning the game was how you got ahead in America, just as generations of people from different backgrounds had done before. You did not fuss about work, you just got after it. Other engineers I've spoken to had similar experiences in the 1970s, when offshore drilling businesses were concentrated mostly in Houston.

In due course, I would join the Society of Petroleum Engineers (SPE) to learn more about the petroleum industry and to learn to fit into this new society. It was two or three decades before SPE invented Young Professionals and developed special programs to cater to the sensitivities of new hands finding their way in the oilpatch.

It's true that oilpatch culture has definitely shifted from what it once was. When I started, things were much more rough-and-tumble, much less politically correct. Aggie jokes were popular, just like how we made funny jabs in Scotland about the English and the Irish, and up north in America it was about the Italians, and Polish. I considered Aggie jokes somewhat unfair, as all the Aggies I met were pretty good engineers.

Today, many of the colorful expressions are no longer heard. In a conference planning meeting in 2014, I suggested a weevil school for Young Professionals as an alternative to SPE's roughneck camp but quickly discovered that no one knew what a weevil was.

In 1967, when Houston newspapers and TV commentators spoke about "snowflakes," they meant that stuff that came from the sky in very small quantities maybe every third year in Houston, and quickly became wet drips— quite different from the snowflakes they and cable news talk about today!

The Start Of ETA (1970)

ETA started business in the first weeks of 1970.

It was a different time. The Vietnam War was at its peak and Richard Nixon had entered the White House, succeeding Lyndon Baines Johnson of Texas. People bought records (45s or the new LPs) to enjoy the hip and fresh music of The Beach Boys and The Beatles, never thinking their tunes would be classics fifty years later.

In 1968, people in Houston flocked to see "The Hellfighters," a movie based loosely on the lives of Red Adair, "Boots" Hansen, and "Coots" Matthews, who put out horrendous oil well fires in remote parts of the world. It starred John Wayne, tall and brash, as "Red" Adair; this was somewhat amusing to Houstonians, as the real life "Red" Adair was an unassuming guy of medium height who was modest despite his risk-taking (as I can attest, since I met him in a restaurant in what is now the Galleria area).

During July of 1969, Neil Armstrong had walked on the moon and radioed his famous words back to Houston. People still read news magazines, and especially sought out the July 18, 1969 special issue of *Time*, "To the Moon Special Supplement." It carried the normal news topics of the day with a 14-page feature: on one side, a photograph of "Astronaut Armstrong with Lunar Module;" and symbolically on the opposite page, a drawing of "Christopher Columbus Aboard the Santa Maria." It was historic major news like never before! America was great.

The following month, in August 1969, the first Woodstock festival kicked off in the Catskills in New York state. It was a time of change, when anything felt possible.

In 1969, I drove to Austin, Texas to appear before the Texas Board of Professional Engineers. I had a mission: to convince the board that my Master of Applied Mechanics degree from University of Virginia in 1964 deserved accreditation in Texas, even though that particular degree was not on their approved list. I needed this accreditation for registration as a professional engineer. The Texas Board investigated properly in the most professional manner and approved me.

So, in 1970, three years after arriving in Houston from Scotland, there I was, trying to build a business in the petroleum industry— 29 years old, new to the world of business in America, and just beginning to understand the good ol' boys of Texas.

I had found an opportunity to provide consulting services in structural mechanics, using a combination of new technology: publicly available computer services and existing structural analysis programs. I had met Ed Lowery who said he was an engineering graduate from North Carolina State University, with an MBA in General Management from University of Southern California. He was attempting to build a management consulting practice.

He had an impressive credentials–his resume read, "Ed received three B.S. degrees

(Electrical Engineering, Engineering Mathematics, and Engineering Physics) from North Carolina State University at Raleigh. He did graduate study in business at Rutgers University, and graduated top of his class at University of Southern California with an MBA in General Management."

In 1970 I thought his background in both engineering and business management was an excellent complement to my history in engineering and the talents I felt I had in building and bringing in the business from the prospective client base. So I teamed up 50:50 with Ed Lowery in starting Engineering Technology Analysts, Inc. (ETA), where he served as business manager, dealing with administrative and financial matters.

While fact checking this book in 2018 that line up of academic credentials sounded a little unlikely, rather extreme. I tried to verify Ed Lowery's degrees, and according to www.degreeverify.org found that North Carolina State University at Raleigh had no record of attendance by Edwin L. Lowery or his being awarded any engineering degrees! The same verdict came back from University of Southern California: no record of attendance by Edwin L. Lowery or his being awarded any degree. The data service cautions that people may choose to seal their records but with such apparently stellar academic histories, why would anyone want to do that?

It would have been a warning sign for future events. And I heard comments later about these 50:50 ownership deals often being unwise. But back in 1970 I took people's word on important things like academic records and charged ahead with youthful enthusiasm.

ETA started in one of the very modest, low-rent offices at 3310 Richmond Avenue in Houston, a two-story building that has long since been torn down.

As president of the company, I took on the responsibilities of chief engineer, a title which sounded good to an enterprising, young engineer, but really meant doing everything; I was "chief cook and bottle washer" as the saying was. Still, I was the one with the Texas Board-approved professional engineer stamp number 29619, endorsing that everything ETA did was sound. I was the engineering analyst tackling whatever work that a client wanted us to do. Whatever the task, I figured out how to model the assignment using my knowledge of structural mechanics and the computers and software available in 1970.

I wasn't going into this venture completely blind. It helped that I had prior experience during 1965-1966 using an IBM 7094 in the structural analysis and design of a 152-ft.-high geodetic structure in UK and knew some FORTRAN.

It helped that my partner's wife, Janice, was a programmer and could create software to set up the loadings I'd formulate. Then we had to automate design code checks.

The Stone Age: State-Of-The-Art Computing For Houston Engineers (1970)

When ETA started in 1970, there were many companies competing to provide different computer services. This new and complicated upheaval of old methods was confusing to many engineers, who were anxious to simply get their engineering work done. They did not want to spend time learning the ins and outs of cutting-edge and diverse computer systems or the intricacies of engineering analysis programs. As a new business in the realm, we navigated our way through a bewildering scope of computer services.

Table A gives an idea of the range of computer service companies available in 1970. There was one company with very bright engineers that offered services on an analogue computer, apparently used for process designs by local companies. All the rest were digital. Fourteen timesharing computer services had entered the market in Houston, offering a concept that had quickly become popular. It used an acoustic coupler to connect by phone with a remote computer. Input and output clacked out on a teletype printer. A decade later, they all had just about disappeared or the companies had morphed into something else.

Banks and oil companies owned and operated mainframe computers for their administration reasons and sometimes would also rent them out. Service bureaus owned and managed large computers that were able to handle the structural analyses we wanted to do. In some cases, these bureaus had software packages.

Engineering calculations on computer meant developing special skills, using data centers with millions of dollars in Univac 1108 or CDC 6600 equipment, and employing a staff to care for them. Or there was another option: hiring specialists from a company like ETA.

Analog/hybrid systems:

KinOTrol, Inc.   DEC PDP 10/50 - SS100

Digital Systems -- Batch:

(a) Large Scale:

    Computer Knowledge Corporation   CDC 6400
    Control Data Corporation   CDC 6600, CDC 3600
    McDonnell Automation   CDC 6400, IBM 360/85
    University Computing   Univac 1108

(b) Medium Scale:

    Commercial service bureaus   ⎫   IBM 360/20
    (refer to yellow pages, Data   ⎬   to
       Processing)   ⎭   IBM 360/50
    Excess capacity at banks,   IBM 1130
    oil companies, consulting   Univac 9200, 9300
    engineering companies   IBM 360/20 to 360/65

Digital Systems -- Timesharing:

Academy Computing Corporation   GE 255, GE 430
Allen Babcock   IBM 360/50
Computer Complex   XDS 940
Computer Applications   TSS8 (DEC PDP 8)
Corporate Computing Inc.   XDS Sigma 7
General Electric Co.   GE 265, GE 635
Honeywell Information Services   H 1648
ITT Data Services   IBM 360/65
McDonnell Automation   XDS Sigma 7
Service Bureau Corporation   IBM 360/50
*TIMAR, Inc.   DEC PDP 10
Tymshare Inc.   XDS 940, XDS Sigma 7
United Computing Systems   CDC 6400, GE 265
University Computing Co.   Univac 1108/PDP 8/PDP 9

*Works only with large volume sophisticated users.

Table A:

Computer Services available in Houston for engineering calculations in 1970.
Source: <u>Houston Engineering Science Society (HESS)</u>, *Slide Rule, November 1970*

All these computer service companies would have salesmen chasing business in what must have been a pretty limited market, but they were driven by the dream of what computers could do. And they were right—to a fault. With the rate that technology advanced, I would guess that every one of these companies in Table A aren't in the market today!

The ETA Name And "Efficiency Through Computing" Logo

Our company name of "Engineering Technology Analysts" was a straightforward and succinct statement of our founding goal.

The ETA logo showed the company name with punched paper tape below it. Punched paper tape was one of the methods used to input data into computers and telex machines to communicate with remote places in the oil and gas world such as Singapore.

Superimposed on the strip of punched paper was the Greek letter eta (η). It was the Greek letter that stood for efficiency in thermodynamics, as I had learned from my classes at University of Glasgow. At the time, we were proud of our streak of marketing genius in creating our logo, though I doubt anyone ever really paid much attention to it. You can see the logo at the top of *ETA Innovation*, the quarterly newsletter we published, shown in Picture 1.

During a vacation to Greece in 1971, I had a rug made that was emblazoned with an η. I kept it in my office after we grew, even when, in 1972, we moved to occupy the third floor at 4140 Southwest Freeway in Houston.

In 1975, only a few years later, we changed our name to reflect that we had moved away from being an analytical firm to be more of a full-service engineering and design firm; thus Engineering Technology Analysts Inc. became ETA Engineers Inc.

Engineering
Technology
Analysts, Inc.

INNOVATION

A privately circulated engineering review

Volume 1, No. 1
First Quarter 1972

Why not a Jack-up for 400 ft. water?
New developments in piping analysis

Picture 1:
Front cover of the first issue of *ETA Innovation*, illustrating the ETA Deepwater Jackup, a slant leg design for 400-ft. water depth.

Starting Out With Pipe Stress And Structural Analyses

At first, ETA did a lot of pipe stress analyses, taking account of dead weight, temperature and pressure effects in complicated piping systems with many branches, bends, valves, and connections to equipment. Sometimes we worked for engineering contractors such as Brown & Root that were building petrochemical installations. At other times, we worked for gas transmission companies such as Columbia Gas Transmission, El Paso Natural Gas, and Trans Canada Pipelines that were building compressor stations in different parts of North America.

Diagram 1 illustrates the type of analysis problem we had to investigate to ensure that piping systems were in compliance with ASME B31.8 for gas transmission systems. For safe reliable operation of the compressors, it was critical that the piping did not impose excessive forces that might distort compressor casings and affect smooth operation (hence the re-run indicated in Diagram 1 to confirm a safe piping configuration).

We used the publically available Mare Island Pipe Stress Analysis program, a complicated and cumbersome piece of software originally developed by the U.S. Navy in its Mare Island Naval Yard near San Francisco, California.

This was an era of using punched cards for data input, checking and re-checking the punched cards to get the data right, and then using a nearby data center with a Univac 1108 computer— or the CDC 6600 at a data center miles away on the Gulf Freeway. We simplified the output through writing a post processor to compare computer results against ASME B31.3 and B31.8 so the results were more useful to our clients.

At one point, we had to figure out how to do big deflection ("nonlinear") analyses for the Trans-Alaska Pipeline System (TAPS). A section of the pipeline slid over its supports above the tundra, as it expanded and contracted because of the hot oil flowing through it. This area was also subject to earthquake loadings. Because of the sliding ability, forces could be computed up to a certain level. Then sliding would occur above that level, changing the force and stress distribution. TAPS engineers were worried how far it would slide and whether the pipeline would slide itself off its supports due to either expansion and contraction or an earthquake.

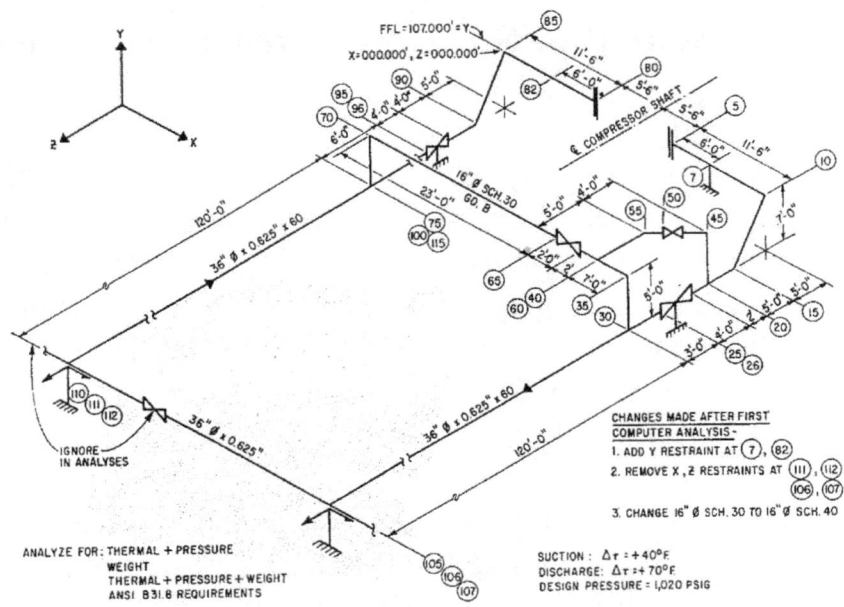

Fig. 1—Sample problem, where a complete stress analysis is required to meet ANSI B31.8 and manufacturer's loading criteria at compressor flanges. The system is shown as it was initially designed and changes recommended after first computer analysis.

Fig. 2—Results from the initial run on the sample problem in a computer output, showing that the forces and moments acting on both compressor flanges are too high.

Diagram 1:

Pipe stress analysis: Typical problem and results

Source: "Computer Program Checks Station/Piping System Design," <u>Pipe Line Industry</u> *February 1971, pp. 88-91*

We quickly devised a way to use the standard Mare Island program to iteratively solve this unusual nonlinear analysis problem. Before us, the professors at Texas A&M had spent months on this issue and had been unable to model it! We concluded that we were really smart.

We branched out in the assignments we took on. At The Offshore Company (TOC), I had learned something about the structural behavior and loadings on jackup offshore drilling units. So ETA performed computer-based structural analyses of jackups. Our first client for the structural analyses of jackup legs was LeTourneau Offshore in Houston. We prepared a proposal for our services dated November 23, 1970, and I visited Ken Farmer (President), Searcy Birdsong (vice president of finance & administration) and Harry Campbell (chief engineer) at their offices in a nearby office on Kirby Drive. They were a friendly and professional group, and before long we started doing assignments for LeTourneau Offshore on their 52 and 53 Class jackups. It was a good working relationship.

Again, there were long computer runs at the data centers, often at night. We had to figure out what loads we should really apply to simulate wind, wave, and current, along with the operational loads for the jackup's legs. We had to develop the right structural model to analyze.

And then there were simulations of loadings on the jackup leg while the jackup was afloat and under field or ocean tow, swinging from side to side, to determine whether leg stresses became high enough to demand removing upper portions of the legs for these tows.

While ETA made a name for using the latest in computer methods, our engineering practice still demanded both basic and long calculation books. These books outlined the equations and theories behind the engineering principles we employed, so they could be checked by third-party regulators or class societies if need be.

These documents were usually a compendium of work by two or three engineers. Each engineer tackled different aspects, such as basic structural analysis of the legs and hull, drawing from both manual and computer-based methods.

Then there were assessments of the loadings while the jackup was rolling during a wet tow. We had to divine what would be the most extreme loadings on the legs and on the spud tanks. The pages reproduced in Diagram 2 are typical of such assessments, with different handwritten work coming from different engineers. It is an engineering analysis of a LeTourneau 53 Class jackup, which was a the latest in jackups in 1971.

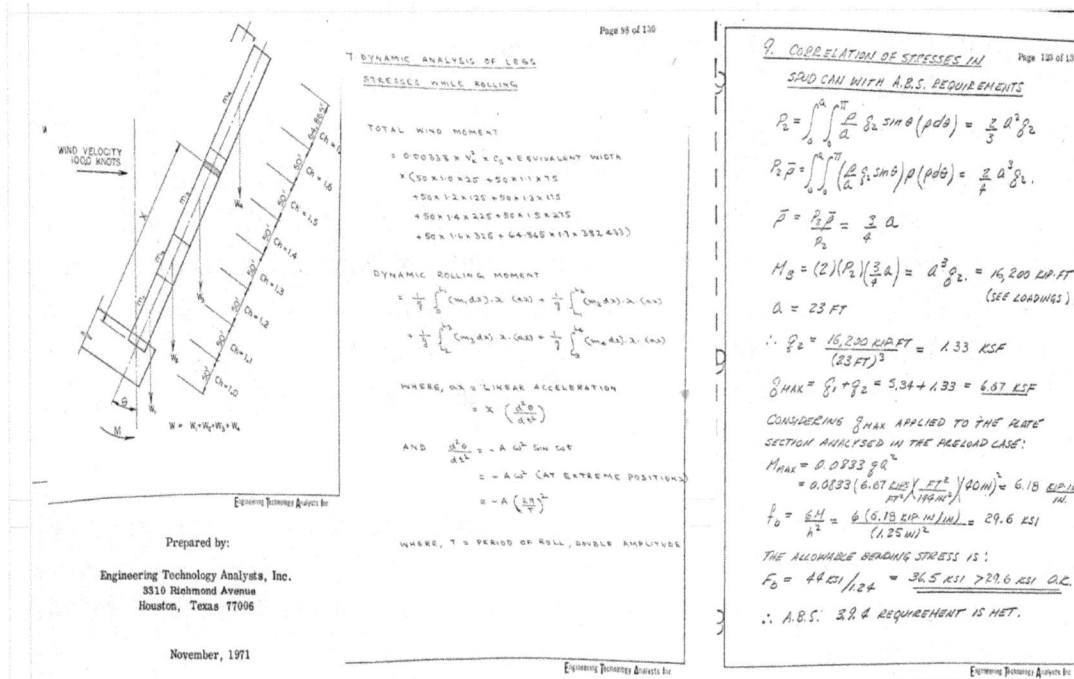

Diagram 2:
Handwritten calculations on a LeTourneau 53 Class jackup
Source: <u>ETA calculation book</u>, *November 1971*

This kind of work stimulated the engineers working at ETA; they had spent years on advanced degrees to learn skills they were now able to apply in serious real life work.

One of our jobs was investigating the leg strength of *Penrod 55*. Afloat with its 467-ft. legs (a record leg length at the time), there were concerns about the stability of the unit and the strength of the legs during a field move, when they were subjected to rolling. *Penrod 55* was being moved in the Gulf of Mexico, quite often with 50 ft. of leg below the hull because of these potential issues.

This was before dry tows moved jackups onboard heavy-lift vessels. The motions during wet tows could often impose greater loads on the legs than today's dry tows.

Wet tows were demanding on the floating stability of the jackup, and we had to assess all the effects.

A heavy-lift vessel has the ability to navigate around extreme storms and travel perhaps three to four times faster than a jackup in a wet tow. The wet tows of the 1970s were a particularly serious risk exposure to be dealt with, calling for substantial effort in often state-of-the-art and controversial engineering analyses.

Some clients would ask us to perform analyses for intact and damaged floating stability for roughly triangular-shaped jackup hulls (the geometry was relatively straightforward but seriously differed from the ship-shaped hulls that naval architects were used to investigating).

Analyses got a lot trickier after Forex Neptune of France contracted with LeTourneau to build one of their Pentagone designs of semisubmersibles at their yard in Brownsville, Texas. The configuration of the *Pentagone 82* was far from straightforward: a pentagonal configuration with five vertical columns, each with a round hull at the bottom, connected with large buoyant braces. It was a very difficult configuration for normal stability calculations— which is why our client (LeTourneau) wanted someone outside their company to do it!

Rather than use manual graphic methods for the calculations, we adapted U.S. government-developed software written in Fortran. It was in the public domain and had originally been made for much simpler barge and ship-shaped hulls. We had to adapt it to use on the unusual Mobile Offshore Drilling Unit shapes of hulls for our hydrostatics and in our intact and damaged stability work.

These methods meant more late-night, brain-stretching and debugging before we were able to figure out the results needed for the odd shapes.

Detecting A Market Need For Deepwater Jackups

During 1970-1971, ETA completed many computer-based structural analyses of jackup legs for LeTourneau, Zapata, and others. We learned what the available designs of jackup structures could do in various situations: on location, during extreme storms, as they got on and off location, and during field moves and ocean tows.

Towards the end of 1971, as we learned about jackups and the drilling market they served, it occurred to us that (quite separate from our engineering analysis practice) there might be a real need for new jackups that could be designed to work beyond the typical limit of 300 ft. of water which was truly "deepwater" back in 1971!

The rationale was simple. Semisubmersibles cost more to build than jackups. In addition, in 1970, semisubmersibles often worked in similar or slightly more water depths, as indicated in Diagram 3.

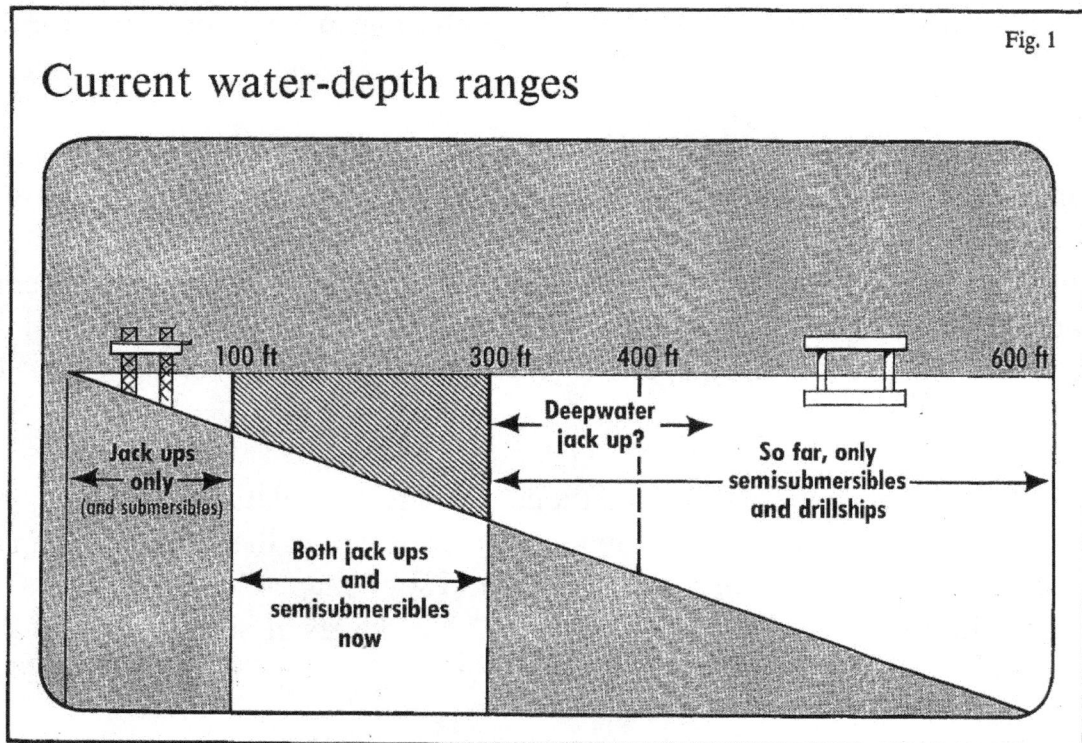

Diagram 3:
Market segments for jackups and semisubmersibles in 1971-1972
Source: ETA article in Oil & Gas Journal, *North Sea Report, January 10, 1972*

The jackup was simpler to operate and had a little more uptime. Separate from ETA, drilling contractors had observed the basic need for more deepwater capability and had started to order more deepwater jackups and more semisubmersibles, all as indicated in the trends in Diagram 4.

Diagram 4 also shows the trend in the growth of business as the number of jackups and semisubmersibles under construction increased. The market for jackups seemed to be coming together with ETA's design dreams!

If ETA could devise a jackup design for, say, 400 ft. of water, we would really be onto something!

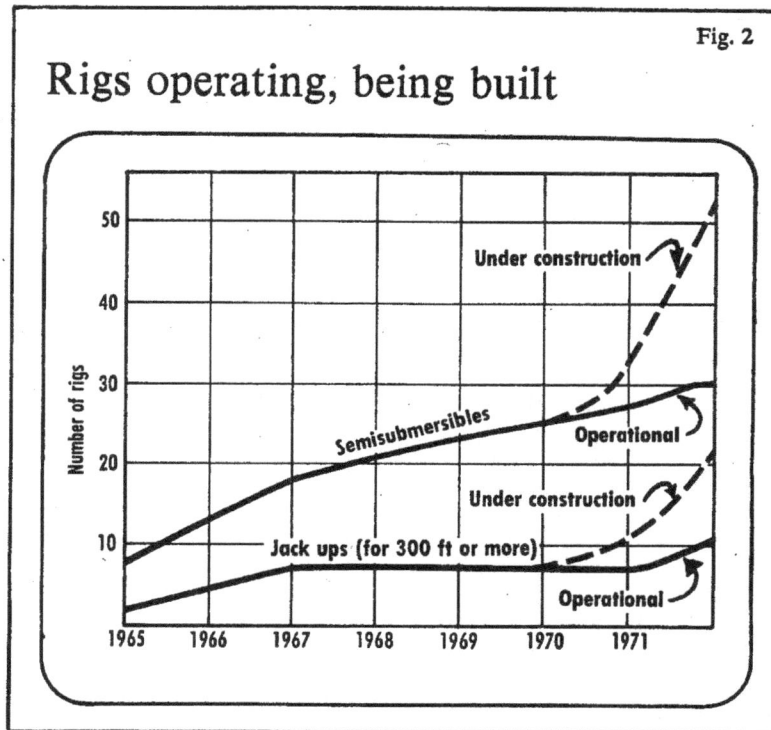

Diagram 4:
Jackup construction trend for 300 ft. w.d. jackups, 1965-1972
Source: ETA article in <u>Oil & Gas Journal</u>, *North Sea Report, January 10, 1972*

ETA's First Jackup Design Concept: A Slant Leg Jackup For 400 ft. Of Water (1972)

I first learned about slant leg jackups in early 1971 while visiting Dixilyn Corporation in Houston to explore consulting business, perhaps on leg strength analysis for their jackups. Their *Dixilyn 250* was able to work in up to 250 ft. of water with three truss legs that were 376 ft. long. LeTourneau Offshore built *Dixilyn 250* at their Vicksburg yard in Mississippi and delivered it in September of 1963. It had been a bold move to build this jackup and put it to work in the Gulf of Mexico!

The office and the president were just as impressive. I met with the company president and his marketing officer in a large, opulent office in downtown Houston. The gigantic, traditional, dark wooden desk glistened and had a black rubber mat on the floor at one end. In the middle of the mat sat a big, shiny spittoon. The president wore boots and a business suit and chewed on a huge, unlit cigar.

ETA never did any business there, but it left an impression and I learned a bit more about offshore drillers.

It was about then that I learned about what it really meant when someone said they were "just an ol' country boy from East Texas." It was really a warning signal (usually not a good one) to listen very carefully to what came next. The comment was designed to put you at ease and lay the groundwork for a proposition. That proposition could be a genuine offer of an alternative course of action, but more often it would be a one-sided, convoluted, or unacceptable idea.

Part of life in living in Texas was that immigrants and other newcomers from up north also had to learn the concept of "all hat and no cattle." This phrase could have different shades of meaning. It referred to someone or something that had little substance and was mostly just talk. But it also was a euphemism for outright deception and Texas bullshit.

We did secure another consulting assignment in 1971-1972, working for Scott Kobus, an industry pioneer with Zapata Offshore, one of the few drilling contractors which pioneered the use of jackups from the early days. Although I was unaware at the time, Zapata had been started by a George H.W. Bush. It was years before I properly understood his pioneering significance as Chairman of the Board of Zapata Offshore and the respect his name earned as he progressed in elective office.

Kobus wanted some computer-based structural analyses of their new slant leg *Zapata Nordic* jackup. After getting acquainted with Zapata and their slant leg jackups, it struck me how the slant leg configuration was indeed a very efficient structure overall. Despite the added complexity of the legs going through a frame hinged at the hull, the connection details, and the need to use a locking mechanism, the slant leg configuration offered an important advantage; in late 1971, I saw the potential for designing jackups for much deeper waters, up to 400 ft. This depth sounded insane to many because 300 ft. was testing the limits in that day.

That engineering concept coincided with the market potential shown in Diagrams 3 and 4. We thought there was some real merit to our new design, so we wrote up the vision for our ETA Deepwater Jackup, a slant leg jackup. We put an artistic impression of this idea on the cover of the inaugural issue of *ETA Innovation*, which we published in the first quarter of 1972 (shown in Picture 1).

Our article in the January 1972 issue of *Oil & Gas Journal* had alluded to the idea: "Jack up for 400 ft. water depths feasible" but that previous issue did not show what it really looked like. Illustrations were not published until April 1974 in *Offshore* and *Northern Offshore*. We talked to potential clients about our jackup ideas, and the word was getting around that ETA was pushing the limits on jackup designs. It was a deliberate marketing ploy.

Brainstorming led to feasibility calculations that looked promising, so we sought to develop this idea into a design that could actually be built.

Our good client—and the largest client for jackup work— was LeTourneau Offshore, which by then had been acquired by the Marathon group and was now known as Marathon LeTourneau Offshore Company. We made our first proposal dated November 7, 1972, offering to develop our deepwater jackup design for their use. We provided a package of fourteen drawings and a summary specification with design criteria to show expected performance.

We presented and discussed our 12-page proposal with the drawings package in a meeting with Mr. "Johnny" Woods, the vice president of engineering. Our proposal was met with interest and polite curiosity, but that was about as far as it went. It was probably as we should have expected; we had had the audacity to suggest a design could be stretched to 500 ft. water depths with 695-ft.-long legs. Hearing this in 1972 from a bunch of young guys, they must have thought that in ETA we were smoking something strange. There's more about this design later in Part IV, "Brainstorms, concepts, and patents: Paper only! The jackup with one leg that inspired others."

.

As time moved on, we did more and more consulting assignments and chose to develop and refine our analytical tools to compute wind, wave, current, and other loadings on offshore structures. For example, back in 1972, it was a relatively big deal to formulate Stoke's Fifth Order theories to generate wave and current loadings to apply in computer-based structural analysis methods. That new capability had led to this improbable brainstorm of a jackup for 400 ft. of water that could really be backed up with calculations. Software was made to check leg structures against design codes such as the ABS MODU Rules that were the new standard, published for the first time in 1968, just three years before our analyses.

Challenging The Establishment Of The Day: The Original Big Four

In retrospect, ETA did not fully recognize the magnitude of the challenge faced in competing with designs from proven jackup designers. Nor did we truly realize the extent of the long, proven experience of the jackup builders. But with youthful enthusiasm, we charged ahead and did it anyway.

Table B summarizes the jackup deliveries during the six years of 1970-1975 from the Original Big Four. There was an industry preference towards rack and pinion elevating systems over the hydraulic jack and pin systems; this is evident from Levingston Shipbuilding and Marathon LeTourneau's delivery of a total of 23+5=29 jackups in contrast with the total of 9+3=12 employing the Bethlehem and The Offshore Company hydraulic jacks and pins systems.

The preference for rack and pinion elevating systems would increase in the subsequent 10-11 years. In addition, preferences in types of rigs would also shift from slot rigs (like the jackups tabulated above) to cantilever rigs.

In 1972, ETA faced four established jackup design sources: three jackup designer-builders (Bethlehem, Levingston and LeTourneau) and one jackup owner, a drilling contractor which designed its own jackups and sometimes licensed out its designs and jacks (The Offshore Company, later known as Sonat and now as Transocean). The Offshore Company also built several jackups in Scotland.

Jackup supplier	Location	Role in jackup business	Jacking system		Jackups delivered 1970-1975
			Type	Supplier	
Bethlehem Steel	Beaumont TX	Designer, builder	Jacks & pins	Self (proprietary)	9
Levingston Shipbuilding	Orange TX	Designer, builder	Rack & pinion	National Supply	5
Marathon LeTourneau	Vicksburg MS	Designer, builder	Rack & pinion	Self (proprietary)	23
The Offshore Company	Houston TX	Drilling contractor	Jacks & pins	Self (proprietary)	3

Table B:
The establishment: The original Big Four jackup suppliers in 1970-1975

Jackups were built in either yards in the U.S. or overseas. Bethlehem and LeTourneau had yards in Singapore that they set up and managed. Levingston Shipbuilding of Orange, Texas licensed Mitsui in Japan.

ETA also faced another important and challenging element to the jackup design business: the supply of the elevating system. Bethlehem, Marathon LeTourneau, and The Offshore Company each had their own proprietary elevating system.

Both The Offshore Company and Bethlehem Steel used a hydraulically driven pin and jack system which they claimed was continuous like the operation of competing rack and pinon elevating systems. Levingston and LeTourneau both used rack and pinion elevating systems that were electrically driven and were clearly continuous and well-suited to deepwater truss leg jackups like the kind we had in mind.

It was felt desirable for a jacking system to be continuous because if one leg went down more than the others, keeping the hull level would be easier. Further, a hydraulic jacking system at mid-stroke meant locking of the leg had to be available, whereas the continuous rack and pinion system was always engaged whether powered on or off. Despite the debates, the operating history with both types was excellent. A new designer could not compete without a supplier for the jacks. In 1972, ETA had only one possibility to take a position in that business.

National Supply was a leading supplier of drilling and related equipment for jackups, and it also manufactured a rack and pinion elevating system. ETA had seen that in 1969-1972 the only designer-builder of jackups that used the National Supply jacks was Levingston Shipbuilding in Orange, Texas. National Supply wanted to encourage other jackup designers and builders throughout the world to buy their equipment and so we met with National Supply and explained the need for an elevating system for ETA's new designs. National Supply agreed to make their elevating system available for use in ETA's new designs.

The jackups we had in mind would also require large quantities of the special five-inch-thick high-yield steel for rack material. Armco Steel was willing to make that high-strength rack material available to match their elevating systems to the strength properties ETA had specified.

These two big advances meant that ETA was able to proceed with the design of its jackups and make viable offers to the market. The way was open now for ETA to compete with the original Big Four with designs such as ETA's Robray 300 Class and its other subsequent designs.

Accidents With Jackups Were Relatively Frequent (Early 1970s)

In the early 1970s, drilling offshore was a new and growing business, still finding its way as water depths became greater and greater and new regions in the world had to be explored. Pioneers had to take on responsibilities for creating new designs that would have crews onboard and have to be safe and able to endure serious storms.

As the industry's fleet of jackups grew, accidents were relatively frequent. However, the industry as a whole learned remarkably quickly, and history later showed how accident rates went down during1970-1981 as the tumultuous growth years progressed.

Table C shows accidents for the different types of Mobile Offshore Drilling Units (MODUs). It is extracted from a 1985 paper by two prominent investigators of the *Ocean Ranger* disaster. In the accident, 84 lives were lost when the *Ocean Ranger* semisubmersible went down on February 15, 1982 offshore Eastern Canada. The table shows how jackups had the most operating experience: 1974 rig-years for jackups versus 964 rig-years for semisubmersibles in the same time period of 1970-1981. Drillships and barges had 816 rig-years.

Accident Frequency and Loss Ratios Worldwide

	JU	SS	DS, DB	Total
Rig-years	1974	964	816	3754
Accidents	135	86	104	325
Total losses	24	2	4	30
Accident frequency (accidents/100 rig-years)	6.84	8.92	12.75	8.66
Loss rate (total losses/100 rig-years)	1.22	0.21	0.49	0.80

Major Offshore Accidents by Degree of Structural Loss and Different Rig Types

	JU	SS	DS, DB	Total
No damage	15	27	10	52
Minor damage	26	32	38	96
Damage	42	18	39	99
Severe damage	25	6	11	42
Total loss	24	2	4	30
Unknown	3	1	2	6
Total	135	86	104	325

Accident Frequencies by Type of Rig

	JU			SS			DS, DB			Total		
	Rigs	Accidents	Frequency	Rigs	Accidents	Frequency	Rigs	Accidents	Frequency	Rigs	Accidents	Frequency
1970	91	13	14.29	25	3	12.00	43	7	16.28	159	23	14.49
1971	97	9	9.28	27	5	18.52	50	10	20.00	174	24	13.79
1972	106	8	7.55	31	5	16.13	51	9	17.65	188	22	11.70
1973	117	11	9.40	36	11	30.56	52	9	17.31	205	31	15.12
1974	122	14	11.48	57	11	19.30	56	5	9.93	235	30	12.77
1975	139	10	7.19	80	13	16.25	68	8	11.76	287	31	10.80
1976	163	13	7.98	118	9	7.63	83	13	15.66	364	35	9.62
1977	164	12	7.32	112	2	1.79	90	7	7.78	366	21	5.74
1978	198	7	3.54	121	7	5.79	80	8	10.00	399	22	5.51
1979	221	10	4.52	118	5	4.24	80	9	11.25	419	24	5.73
1980	251	17	6.77	118	10	8.47	79	12	15.19	448	39	8.71
1981	305	11	3.61	121	5	4.13	84	7	9.33	510	23	4.51
Total	1974	135	6.84	964	86	8.92	816	104	12.75	3754	325	8.66

Key: JU = jack-ups; SS = semisubmersibles; DS, DB = drillships, barges.

Table C:
MODU accident data, 1970-1981
Source: Johnson, R.E., Cojeen, H.P.: "An Investigation into the Loss of the Mobile Offshore Drilling Unit Ocean Ranger." Marine Technology, Volume 22, No.23, pp, 109-125, April 1985

Jackups had the most losses at a total of 24, in contrast with 2 for the semisubmersibles and 4 for drillships and barges. It was no wonder that the classification societies, underwriters, and governmental regulators had become concerned in the 1970s about the safety of jackups, and particularly for new designs of jackups from some new upstart outfit like ETA!

Digging a bit deeper one can detect a positive historical trend in the accident data in Table C. Despite the booming growth of the fleet, accident rates decreased significantly (if not smoothly) over the passing years. 1970 had an operating fleet of 81 jackups but suffered 13 accidents that year, a disturbingly high proportion. In 1975, when the design drawings were being delivered to the builder for the *Dyvi Beta* and *Dyvi Gamma*, the industry had more jackups in operation (130), but there were only 10 accidents. This was half the rate in 1970, when ETA started in business.

Later on, 1981 saw even more jackups in operation (305) with 11 accidents, an accident rate a quarter of 1970's. ETA might have liked to say it had a lead role in that downtrend, but the truth was the industry in general managed to do better despite all the growth and many new developments. It matured.

For comparison, the semisubmersible fleet had a worse record in the early years but improved in 1976 onwards, while drillships and barges had the worst percentage of accidents in their fleet.

We did not realize that these improving trends in accident rates were going on while we were in the middle of designing ETA's jackups. In the face of industry pressure from the accident history of 1970-1973, all we felt was the demand to design for fundamentals. Everyone in the industry had to do better. In hindsight, the trend in accident rates going into the 1980s reflects the industry's earnest desire and effort to do just that.

The ETA Robray 300 Class Design (1973)

Robray Offshore Drilling Co. was founded jointly by Robin Loh, a real estate owner and businessman in Singapore, and Ray Williams, who previously worked with Houston's Reading & Bates. In 1973, when they talked to established designers and builders like Levingston, Marathon LeTourneau and Bethlehem, Robray Offshore Drilling Co. could not get the delivery, performance, and terms they desired for the jackups they wanted to own and operate. Robray had heard talk of ETA, seen ETA's published articles, and felt there was a chance they might secure economic and performance advantages over their competitors in the offshore drilling business with a new design from ETA.

So Robray's Ray Williams (president) and Harold Thrower (vice president of operations) came to see me in Houston. They both grilled me on the performance of ETA's jackup designs and asked if ETA could provide a jackup design for their requirements in 300 ft. of water in the Far East.

I went through the high variables capacity we offered:7,400 kips, which blew the typical 3,500 to 4,000 kips offered by competitors out of the water. We discussed the performance in ocean moves with 425-ft.-long legs full up and no leg removal for a motion that was 15 degrees off-center. These moves used the new "ABS Rules for Building and Classing Mobile Offshore Drilling Units" of 1968 and the later version of 1973. I explained how all of this could be done with a lower overall steel weight than the jackups owned by Robray's competitors. .

This was the opportunity ETA was looking for. We feverishly worked up a fresh design proposal, arranging equipment and systems to suit the needs of our prospective client and drawing on what we'd learned from other jackups.

To get this new design completed, we also faced the practical matter of assembling a proposal for the use of ETA's services. ETA had never taken on this amount of engineering before. This proposal had to work for us; we were signing on to a large and demanding workload, which had to be scoped out thoroughly and paid for. The proposal came out to around 71 pages, with lists of drawings we would prepare for a "bid package" to enable binding contracts to be secured by Robray for construction of the Robray 300 Class jackup.

It was important to secure ABS approval of the design, which would require rigorous calculations and many man-hours of high-caliber engineering. We spelled out the team of engineers we would assign and how we could cite man-hours of both engineers and for draftsmen.

At this point in ETA's history, we had not done much design drafting, so we were taking a risk. In the proposal, we laid out bar charts of timelines and tables of man-hours for different disciplines: structural engineers and naval architects, as well as mechanical and electrical engineers. We devised a system for engineering support during the construction and the start of operations. We were confident we could pull off this design and what it was supposed to do, but we were entering uncharted territory!

We knew our overall leg steel weight was lower than in competitive designs, so the construction cost should be attractive. This price was due, in part, to the use of cast steel nodes (joints). These joints reduced stress concentrations at joints of the structural members in the legs and improved fatigue life, thereby reducing cracking at leg joints. We felt the cast steel joints could also simplify fabrication of connections by avoiding saddle end cuts on tubular bracing members. To match the stubs on the joint castings would require simple end cuts, whether at the K joint in the truss legs or at the leg chord connection.

However, in February, Robray asked us to next consider a 220 ft. jackup for well service and production drilling, separate from our initial Robray 300 proposal. So we redid the proposal for this new and smaller design. Discussions with Robray continued for some weeks before the new idea was dropped and attention returned to the Robray 300.

The ETA's proposal was successful. I negotiated an agreement with Robray for this design work and it was signed on March 27, 1973. It was ETA's first jackup design contract.

The design contract with Robray marked ETA's entry into the serious jackup design world. It was a newsworthy event, this new drilling contractor committing to build a state-of-the-art jackup designed for 300 ft. of water… and from a relatively unheard-of engineering company!

The ETA team designed and drafted, manually preparing design drawings on tracing paper on real drawing boards with real pencils. We lived in the days before AutoCAD, so we had a real "drawing office," equipped with drawing tables, D sized tracing paper, electric erasers, and electric pencil sharpeners.

Arrangement drawings were prepared for the hull, two living quarters (one for nationals and one for expats), and legs. We planned the piping and electrical schematics, equipment arrangements. We also provided structural drawings of the hull and leg design details with these special joint pieces and the spud tanks design.

ETA mobilized its team of engineers and added draftsmen for the work. Picture 2 shows some of them at work.

Ramesh Maini, Engineering Analyst, George Danek, Engineering Aide, and Steve Guillory, Drafting Supervisor (l to r) at work on structural design work for the Robray 300.

Picture 2:
ETA staff at work on the ETA Robray 300 Class jackup design
Source: ETA Innovation *of 2nd quarter 1973*

The specification document was truly a cut-and-paste job prepared with typists on IBM Selectric typewriters and careful use of Scotch tape and the Xerox machine. (Microsoft Word was still a generation away!) Unlike today's metaphorical 'cut and paste' on a computer screen, this was real cut and paste! Finally came the stage of combining that specification document with all the drawings to create a full bid package that enabled shipyard planning and estimating.

By this time, the word was out in the industry, so ETA could show an artist's impression of the unit on the front cover of its quarterly engineering review *ETA Innovation*, as shown in Picture 3 and Picture 4.

More engineering followed during construction, leading ultimately to the delivery and operation of a series of ETA Robray 300 Class jackups. When we delivered the ETA Robray 300 design package to Robray as agreed in 1973, events that we had not anticipated occurred. Our client Robray Offshore Drilling and their jackup building yard (Robin Shipyard) were somewhat secretive…or at least we did not hear very much back in Houston about what they were up to! In hindsight, we should have tried to keep more in the loop.

It turned out that Robray Offshore Drilling chose to get their first two jackups (*Ednastar* and *Ednarina*) built by an experienced jackup builder, Hitachi in Japan, instead of at their rig building startup affiliate of Robin Shipyard in Singapore.

Almost three years after the design work was completed at ETA, Hitachi delivered the first two ETA Robray 300 Class jackups at their yard in Innoshima, Japan in 1976, shown here in operation in Picture 5 are the *Ednastar* and *Ednarina*. They were later sold to COSL and renamed *COSL 935* and *Bohai IV* and now operate in China.

Industry databases on the world's MODU fleet show Hitachi also built two more ETA Robray 300 Class jackups, the *Nan Hai III* and the *Nan Ha IV*, both delivered to China in 1982.

At the time, it puzzled us in ETA how Robin Loh could steer clear of U.S. restrictions on doing business with both the People's Republic of China and with North Korea. How was he able to get away with selling a state of the art jackup of 300 ft.-w.d.-capability, loaded with the latest in U.S. drilling equipment? In 2017, I learned from Singapore that the U.S. had been more concerned about the sale of technology to Russia back then than worrying about what went to the Peoples' Republic of China or North Korea. But it's still curious what went on in the layers of dealings for such transactions.

Engineering
Technology
Analysts, Inc.

INNOVATION

A privately circulated engineering review

Volume 2, No. 2
Second Quarter 1973

THE ROBRAY 300—THE FIRST OF A NEW GENERATION OF JACK-UPS TO BE BUILT

HOW TO SOLVE CENTRIFUGAL COMPRESSOR FLANGE LOADING PROBLEMS

HIGH PRESSURE PIPING SYSTEMS PRESENT STRESS, FLEXIBILITY CONNECTION PROBLEMS

Picture 3:
Artist's impression of the ETA Robray 300 Class jackup under tow
Source: ETA Innovation of 2nd quarter 1973

The Robray 300 jacked-up for 300-fo
depths.

<u>Picture 4</u>:
Artist's impression of the ETA Robray 300 Class jackup, jacked up and operating
Source: <u>ETA Innovation</u> of 2nd quarter 1973

In the 1970s, rig databases used to show 13 jackups of ETA Robray 300 Class design committed for construction. A 1983 industry study on the worldwide MODU business by Lovie & Co. showed a total of 11. Some ETA jackups that were announced by Robin Shipyards were cancelled as the jackup market deteriorated.

Robin Shipyard in Singapore eventually delivered five ETA Robray 300 Class jackups during 1976-1982 in the middle of the all-time boom cycle in jackup building.

In August 2016, Oilpro (now out of business) ran a story about Robray Offshore

Drilling Co. written by Ian Craven, one of the original people in Robray. It shed light on Robray's startup and their jackups. It illustrated how Robray was really much more of an offshore drilling contractor that concentrated on tender barge drilling operations rather than jackups and how Robray was eventually acquired by Smedvig interests in Norway and the Robray name disappeared.

The potential of a series of ETA Robray 300 Class jackups was thus of more interest to Robin Shipyard than to Robray as a drilling contractor.

Picture 5:
Ednastar and *Ednarina*, the first two ETA Robray 300 Class jackups,
owned and operated by Robray Offshore Drilling Co.
Source: Society of Petroleum Engineers (SPE)

Robin Loh left the shipbuilding and offshore drilling business around 1985. In 1980, near the peak of the jackup building cycle, he had bought 20 square kilometers of grazing land in Queensland, Australia and started the development of a planned community that has been successful and now is home to more than 30,000 people.

Robin House and Robin Way in Singapore are named after him. In 2010, Robin Loh died at 81 on board a flight from Singapore to Hong Kong.

Building The Team

In 1973, as work on the ETA Robray 300 Class design progressed, ETA's payroll grew to be around forty or forty-five people. ETA was fast becoming known as an exciting place to work with interesting cutting-edge assignments.

During 1973-1974, jackups for 300+ ft. of water were a hot topic, much akin to sixth-generation drillships for 10,000-12,000 ft. of water forty year later. The problems in 1973 were much different from those today but the same driving principles got us ambitious young MODU design engineers going!

Most of us in ETA were structural engineers by background, usually with advanced degrees. We had five team members with doctorates. All of us brought the brainpower to adapt and tackle non-structural work such as naval architectural calculations to make sure our jackups would not flip over while afloat! In 1972, we decided to hire a naval architect, Ralph McTaggart, who was a Scot like me.

When Robray wanted someone from ETA's staff to join them and help resolve issues during yard construction, we agreed for Richard Greff from our design team on our engineering staff to go to Singapore.

Experience was treasured and experts sought out to confirm judgment on design choices. Thus it was unusual for a new outfit to secure a design contract like ETA did for the ETA Robray 300 Class jackup.

It was a time when engineers still used slide rules for calculations. Structural and mechanical engineers would usually have books by Timoshenko and Roark at hand in order to find the right formulae to use in calculating stresses and dimensions. And even then the numbers had to match the common sense and experience of seasoned engineers before they were used. I admit to sometimes referring to the textbook from my structural mechanics classes at University of Glasgow, "The Analysis Of Engineering Structures" by Pippard & Baker, which was printed in 1957! It was well before the internet, even before the fax mini-revolution in the 1980s. We communicated only in person or by telex, phone, or airmail. With no smart phones, we actually spoke to one another!

Picture 6:
Peter Lovie at ETA in 1974
Source: "Jackups – A Future in the North Sea,"
interview by June Angerstein, Northern Offshore, *4th*
quarter 1974

About a year later, after news spread that ETA had secured the design commitment for the two jackups for Dyvi Drilling A/S, Norway's *Northern Offshore* magazine did a long interview with me in their 4th quarter 1974 issue. I talked about ETA's projects, designs, and philosophies. In a quote at the end of the article, I said:

> *"We're in an interesting, exciting business and when you have an enthusiastic creative group like this, really functioning as a team, there's very little limit to what you can do."*

As I read this sentence in 2017 for the first time since 1974, I realize these words summed up the spirit of ETA.

Building The Software

The computing of the early '70s was truly primitive compared to computing today;

engineers actually did calculations themselves, and supervisors would verify from their rules of thumb and experience to confirm they were on the right track so that their work could be reliably used in a design.

These computer programs could solve complex matrices of the structural stiffness of three-dimensional structures with six degrees of freedom on each end of each structural member, but there was real skepticism of what might appear in the printout.

It was maybe two decades before PCs were on every engineer's desk. Engineers in the early 1970s were pleased to have a Wang calculator on their desks, plugged into mains power. (The battery powered handhelds came later.)

To give an idea of just how long ago this was, if someone said "Steve Jobs" to a computer person in 1973, he or she might have asked "Who is Steve? Is he looking for people?." "Apple" might have been taken to be symbolically the first computer in human history, the one that Adam and Eve had, which had an extremely limited memory and crashed at the first byte . . .

All of ETA's analyses of wind and wave loadings and of operational weights and loadings had to be combined and fed into third-party structural analysis programs, and the output then checked against design codes (AISC, ABS etc.), usually with post processors we developed to automate laborious hand checking.

Despite all that, the need remained to keep an eye on outputs at each stage to ensure results were realistic. There was no escaping the need for engineering judgment in using these tools— just like today!

In addition to all the structural mechanics, there were floating stability calculations (intact and damaged), hydrostatics, tank tables, and ballasting calculations. All these factors had to fit together. This interdependence called for an ongoing system development effort between our computer programmers and engineers for practicality and usability within the available time and budget.

At the time, there was nothing around to do what we needed, so we had to have two full-time programmers maintain and build the software and help our engineers use it. An October 1974 magazine article in *Petroleum Engineer*, outlined: ETA/WIND, ETA/WAVE, ETA/SPABS, and so on, as shown in Diagram 5.

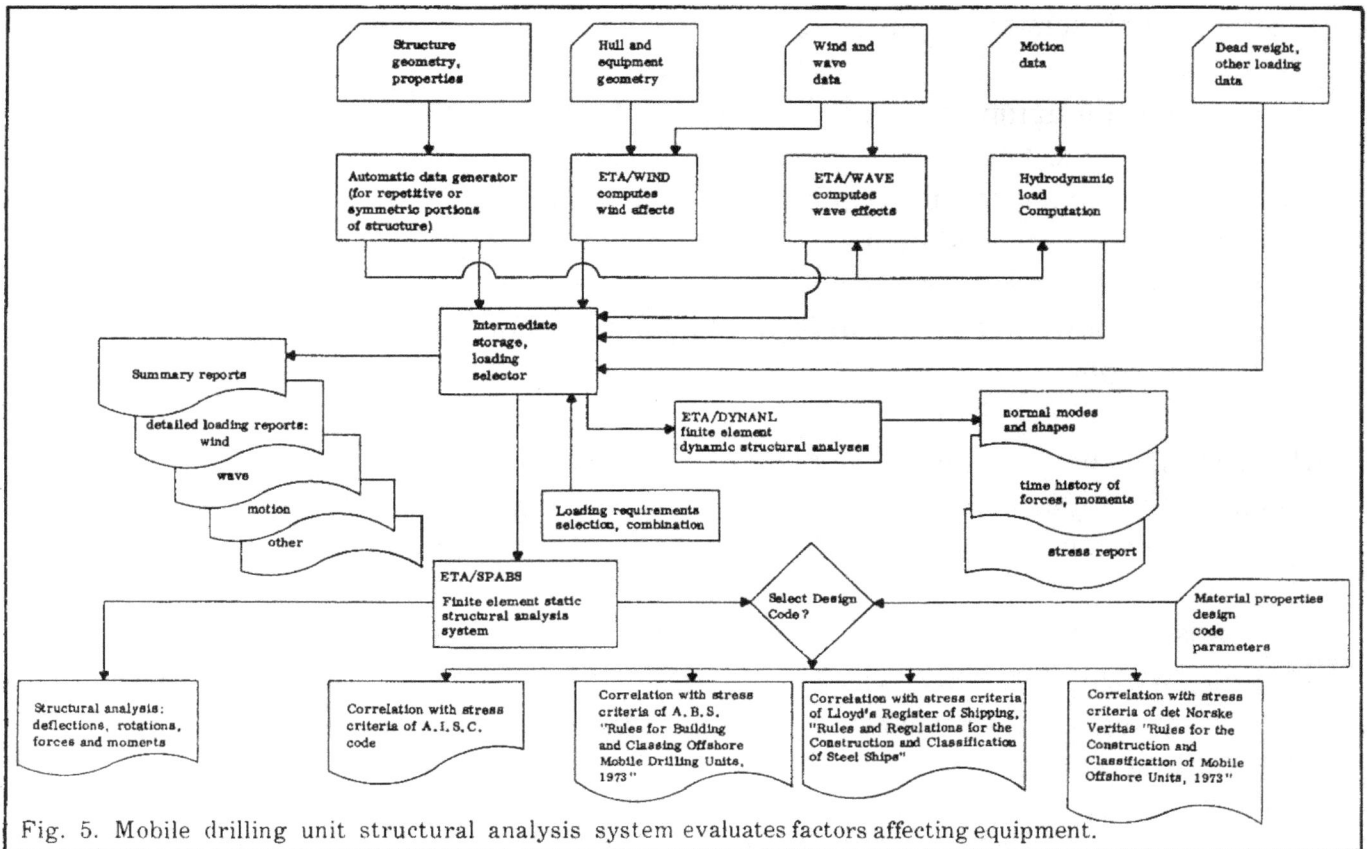

Fig. 5. Mobile drilling unit structural analysis system evaluates factors affecting equipment.

Diagram 5:

ETA's offshore engineering analysis software systems circa 1972-1974

Source: "How Jackups Fit in the North Sea Boom," Petroleum Engineer, *October 1974*

The steady development of analytical tools paved the way for tackling fundamental jackup design and getting fast results. For a fairly small number of analyses—altogether not much more than a hundred or so—these tools were a compromise between front-end investment in programming and getting the work done profitably. Documentation could be brief; it had to be OK for internal use only.

Diversity Before It Was In Style

We had no time to worry about diversity because we had to get the job done with whatever engineers were willing to work and knew what they were doing.

As a young immigrant engineer in Houston, I had seen the unappreciated talents of other immigrants. I had discovered that Houston employers of the early 1970s were sometimes reluctant to hire immigrants, even though they had come to America and graduated with advanced degrees. The unspoken hiring priority I saw was: (1) Texas Aggies and Teasips, (2) graduates from elsewhere in Texas or in the South, (3) graduates from up east, north of the Mason Dixon line or from northern Europe, (4) Chinese, and (5) Indians from India.

Back then, many employers seemed more interested in being comfortable with the potential employee's background, work habits, and university training. In most cases, this manifested in practical reservations rather than any prejudice.

ETA was by necessity a truly equal-opportunity employer, with a diverse team of Indians, Scots, Chinese, black, and white Texans. The chief draftsman's name was Steve Guillory from Louisiana, so we even included Cajuns! It was a time not long after the national civil rights movement. I did not see problems about black and white when it came to hiring. Today questions might be raised about native Americans being considered, or other minorities to be properly inclusive. Life was simpler and we drew on who was available in the marketplace of the day. It all came down to a matter of good engineering, getting along and getting things done.

My openness to the value of hiring immigrants was a common sense move in the marketplace in order to gain first-class talent and brainpower advantages over our competitors. This was how people like Shankarlal Mody, Dr. Rao Guntur, Ramesh Maini, and Carleton Chen came to join us. They wanted a chance to do real engineering and an introduction to the offshore engineering world through ETA.

On one occasion, we had an immigrant engineer with an advanced degree approach us in ETA and explain that he had trouble getting a job and needed to work to show what he could do. I liked his attitude and made up a job for him, a speculation. He turned out to be a very loyal, hardworking, bright, and competent engineer.

The immigrants we hired learned very rapidly and grew as professional engineers, going on to do well in their respective careers in the offshore engineering world long after ETA.

We also had Dr. Doug Bynum, Dr. Dick Gunderson, Dr. Mike Reifel, Dr. Don Smith, and Earney White in ETA's engineering team; after all, we did employ good Texas hands when we could find them!

Between 1970-1975, our team didn't include any women engineers, who were rare in the world and unheard of in the oilpatch. But we did have two Fortran-speaking women programmers (Claudia Leffler and Janice Lowery).

Our diverse team came from our own initiative, not government-imposed regulations, which were not necessary when we valued working with people that took pride in what they did. Our team was spirited and tight-knit.

ETA became known for its end-of-the-work-week parties, which started at 5 p.m. every Friday. When we started in early 1970, these weekly celebrations were one six-pack of beer, but grew MUCH larger in 1974! After one strenuous week, I remember the elevator crammed full of ETA folk heading home after the party, boisterous and somewhat inebriated. They cheered as the elevator brakes slipped slowly and they descended from the third floor, singing en route to the parking lot!

We were intense and young then. As the cliché goes, we worked hard and played hard.

The Floating Stability Investigation

As structural engineers, we knew how our structures would perform and there was plenty of rigorous experience to draw on. Structural analysis software also allowed us to tackle about anything, though it still remained for us to exercise good judgment in modeling the structure and the loads being applied.

In contrast, it was another world altogether to deal with what happened while the jackup was afloat, pitching around in the ocean during field or ocean tows, in calm seas or during violent storms. This business called for more emphasis on historical rules of thumb. As we designed jackups that stretched the limits, we had to make sure they stayed stable while afloat.

In 1973, as we got into floating stability analyses for the ETA Robray 300 Class, we discovered the serious hazard of applying some of the accepted offshore industry rules of thumb (that were based on generations of experience with ships) to the approximately triangular hull shapes we used and our extreme leg lengths of 400+ ft. We ran more analyses, trying to adapt existing experience to satisfy the spirit of the new ABS Rules. As we did so, we uncovered potentially unsafe conditions for damaged stability (one compartment flooded), as well as some difficulty with intact stability.

Fortunately, we identified and exhaustively worked through the problems so that the ETA Robray 300 Class jackups, and others later, were safe for all expected design conditions, though there were some tense months of debate and many, many calculations which taxed our new software and brainpower. It was heated because we were navigating a new frontier; it was a time when jackup stability was controversial and proven, clear guidelines were difficult to find!

ETA chose to put its experience and recommendations in front of the offshore drilling community. In November 1973, ETA's chief naval architect (Ralph McTaggart) and senior engineer (Dr. Richard "Dick" Gunderson) travelled to New York for the Annual Meeting of the Society of Naval Architects & Marine Engineers (SNAME) to present their findings. These principles became a key element in designing the ETA jackups and are further discussed in Part III.

ETA Lands A Contract To Design The World's Biggest Jackups (1974)

Experience with the ETA Robray 300 design led naturally to the creation of other designs: the ETA America, ETA Asia, and ETA Europe jackup designs. From these studies, the ETA Europe Class was the one that went forward into construction for service in the North Sea.

The team of engineers who developed the ETA Europe Class design was basically the same team that had tackled the ETA Robray 300 Class design the year before. We had learned a lot. This time, there were a few additional people and the whole design assignment was seriously more challenging: tougher physical environment, tougher design standards, and a more demanding regulatory approval process.

It was the stuff engineers lived for. And it really did keep us on our toes. The company peaked at about seventy people in 1974, counting the two owners, engineers, programmers, draftsmen, technical writers (Susan Huey and Bonnie Somyak), accountant (Mike Boyle), and several secretaries. The company employed a full-time team, no contractors other than one or two contract draftsmen (e.g. Joe Lovett).

One day, a group of visitors arrived at ETA's offices, led by Mr. Jan Erik Dyvi, chairman of Dyvi Drilling A/S and Mr. Magne Reed, a company officer who had been a UN ambassador for Norway. They were accompanied by Mr. Per Engeset, a broker from R.S. Platou A/S in Oslo, Norway and M. François Goillandeau from Compagnie Française d'Entreprises Métalliques (CFEM) of France. They had lined up potential offshore drilling work, a financing package, and a shipyard construction arrangement and wanted to examine the possibility of building one or more jackups of ETA's Europe Class design.

Over a daylong meeting, I secured a commitment for ETA to start a specification and drawing package to enable firm pricing for the construction of what would become the *Dyvi Beta* and *Dyvi Gamma* jackups.

Dyvi Drilling next retained Det Norske Veritas (DNV) to check our design. After about four weeks of examining our drawings and specifications, DNV said the ETA Europe Class design would indeed perform as we said it would and further; they verified we were right to within one percent (1.0%) on the steel weight needed to meet all their design criteria, something that convinced the savvy Dyvi Drilling A/S to go forward with us. It was an important moment, as it meant that firming up of costs and plans for fabrication could quickly and reliably go forward.

As it turned out, the *Dyvi Beta* and *Dyvi Gamma* would be the first harsh environment jackups for the North Sea and the largest jackups in the world at that time. Built in France by CFEM and delivered in 1976 and 1977, they were then the longest-legged jackups in the world at 508 ft. and the only jackups designed specifically for North Sea service, with a design variables load of a record 9,500 kips, about double what was normal at the time. For the engineers in ETA, designing the biggest jackups in the world was the big league and something we could be proud of!

The American Bureau of Shipping (ABS) had been easy to satisfy on the ETA Robray 300 Class design. I had confided to colleagues that we could have pulled the wool over the eyes of the ABS reviewers. We understood the long lunches commonly taken in the process of getting design changes and new ideas approved by reviewers, but we chose not to BS the ABS; it was more a matter of honor and professional pride in getting designs right by our own standards.

DNV was far more rigorous, but all the designs calculations backed up our predictions for operations on location and during moves. DNV had precedent in their *Rules* for building and classing other hull types, well-developed principles, and a truly rigorous and professional approach. DNV was also engaged in developing their own *Rules for Building and Classing of Mobile Offshore Drilling Units*, and the ETA Europe Class design thus became a test case to test their procedures and philosophies. It was just the kind of challenge that got the engineers in ETA going to rise to the occasion.

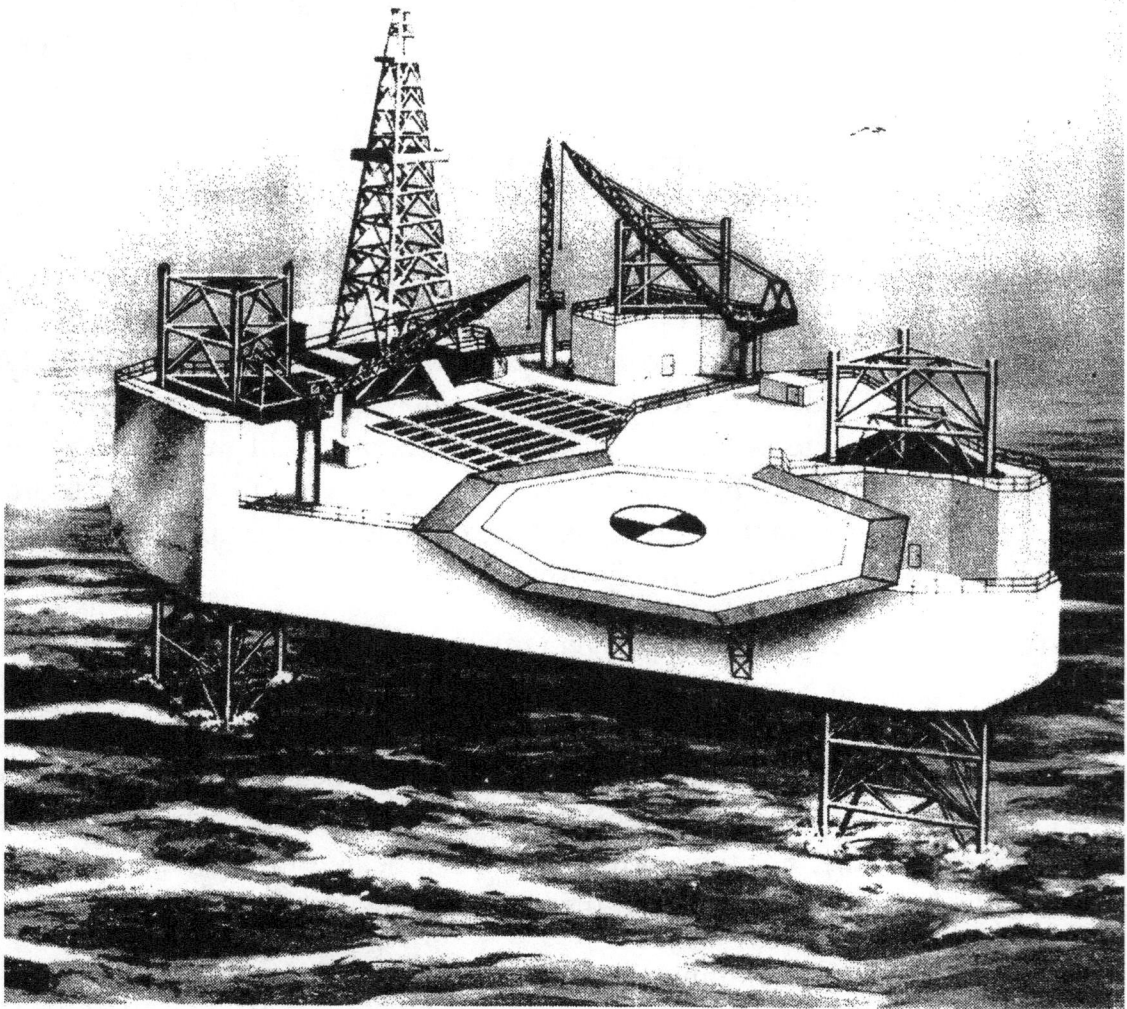

<u>Picture 7</u>:
Artist's impression of the ETA Europe Class jackup
Source: "How Jackups Fit in the North Sea Boom," <u>Petroleum Engineer</u>, *October 1974*

The ETA Europe Class design was the design that everyone in ETA was proudest of. It was truly state-of-the-art, prepared to the highest standards that would work in a demanding physical environment and in a demanding customer and regulatory climate. The design set a precedent for the harsh environment of the North Sea, a category that other designers and builders expanded in the following years. The artist's impression in Picture 7 showed what we expected it to look like.

The attention of DNV on *Dyvi Beta* and *Dyvi Gamma* continued long after delivery (shown in Picture 8) as they investigated and studied the performance of these two jackups in the North Sea during 1977-1985 (further discussed in Part V).

Dyvi Beta and *Dyvi Gamma* were the second and third MODUs in the Dyvi fleet after the *Dyvi Alpha*, which was an Aker H-3 semisubmersible. Like Robray, Dyvi was a new drilling company. (In another coincidence, Smedvig bought Robray in 1986 and acquired Dyvi Drilling A/S in 1988).

The ETA Europe Class design had to be vetted for underwriter approvals for field moves and ocean tows. In the course of this endeavor, I went to see Noble Denton in London for their blessing. They were the independent world authority on jackup moves, whether field moves or long ocean tows with more severe storm exposures. Their office was from another time. Going to this key meeting meant travelling in an old-fashioned elevator with a collapsible lattice door, which was operated by an older man with his dog laying quietly on the elevator floor.

All went well, and I got to know Dr. Tony Denton and Dr. Malcolm Sharples, who confirmed the reliability of the ETA Europe Class design.

Years later, I learned there had been real skepticism toward this new design from this new upstart design company. Noble Denton had gone through our design with a fine-tooth comb but had to agree with this mob of young guys: the design was indeed up to the job.

A decade after that extreme vetting, I worked as an expert witness alongside Tony Denton of Noble Denton on the contentious loss of the *Key Biscayne* jackup in 1983 off Western Australia. The loss of the *Key Biscayne* was a *cause célèbre* that went on for several years, eventually pitting Chevron (who we worked for) against P&O Shipping. It was fun. It took stretching engineering principles and experience, real engineering brainpower and initiative to prepare for this momentous clash! Although I looked forward to visiting Perth, Australia to testify, the case got settled in our client's favor just before trial!

<u>Picture 8</u>:
The two ETA Europe Class jackups on delivery from CFEM
in Dunkerque, France

Source: <u>Society of Petroleum Engineers</u> (SPE)

I still remember a truly beautiful late June evening in 1974 in Oslo Fjord. I was invited to sail on board the yacht of Jan Erik Dyvi with his wife and key associates. After an enjoyable evening, I returned to the hotel, where an urgent message awaited me: "return to Houston, wife in hospital for premature birth." I rushed to the airport for the first flight out the next day– the three-pound preemie survived.

Part V further explores the construction of the *Dyvi Beta* and *Dyvi Gamma* and their subsequent history. Two and a half years after that summer evening, the jackups we were celebrating were delivered, shown here in Picture 8.

ETA Gets Better Known In The Industry

During 1972-1974, various publications wrote pieces about the ETA Robray 300 Class and ETA Europe Class jackup designs; some of these pieces are listed in the Appendix on "Relevant Publications." For example, in 1974, *Northern Offshore* wrote an enthusiastic (if not overstated) piece about the growing popularity of the ETA Europe Class jackups, alluding to the similar rise of the new Aker H-3 semisubmersibles of the day.

It was a deliberate marketing initiative. We could not afford conventional exhibitions, advertising, and business development, so we sought influence via thought-provoking articles. We hoped that management in companies would read these pieces and become potential clients.

Today this strategy might be called thought leadership, but, in reality, we were not so savvy. We just wanted to use our wits for the most bang for the limited bucks in our marketing budget! It worked. Companies that we had never heard of listened. (This is how we had secured the business with Robray Offshore Drilling and Dyvi Drilling A/S.) Steadily ETA became more and more well-known. All the growth and action was heady stuff and we wondered where it might lead.

Doug Bynum (ETA's manager of research engineering) and I studied MODU-ordering patterns. In a moment of weakness, I agreed to publish an article about these future trends. In 1974, there were something like 241 MODUs of all types in operation worldwide, with 130 of them being jackups. *Offshore* magazine assigned a title to our article: "A Thousand Rigs Could be Needed to Meet World Goals."

The piece came out in January 1975, just as newbuild orders stalled and the day rates in the MODU market started to drop. It took only a year or two for it to sink in just how embarrassingly wrong we were! It was a lesson in forecasting anything in the oilpatch.

An Attempt At Diversification: Semisubmersible MODU And Pipelay/ Derrick Barge

While we were heavily focused on jackups, ETA made an attempt to broaden its activities, trying out proposals for the design of other types of floating equipment in a classic attempt at diversification.

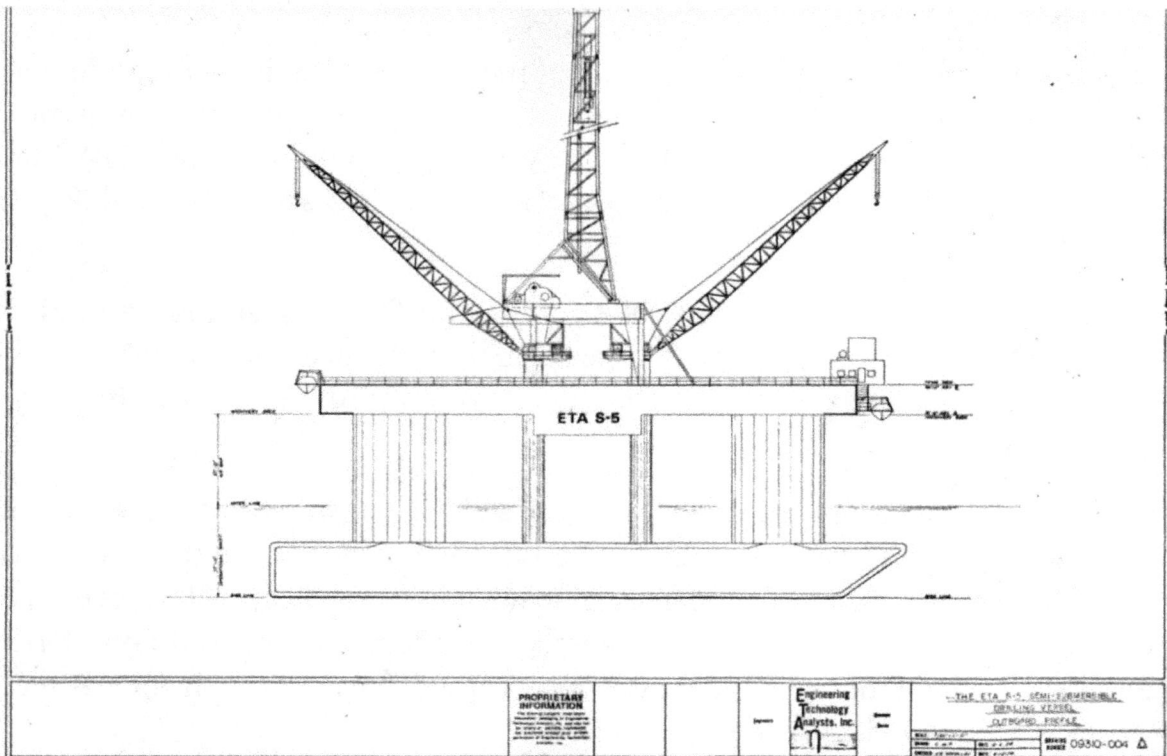

Diagram 6:
ETA's design of compact semisubmersible drilling unit
Source: ETA Drawing no. 09310-004 *dated April 12, 1974*

THE ETA DERRICK/ PIPELAY BARGE

PRINCIPAL CHARACTERISTICS

Dimensions
Length overall . . . . 400 ft. *(122 m.)*
Beam overall . . . . .100 ft. *(30.4 m.)*
Depth . . . . 28 ft. *(8.5 m.)*

Quarters . . . . .220 persons

Heliport . . . . Sized for Sikorsky S-61

Pipelaying Capacity
Pipe diameter . . . . 48 in. O.D. x 0.625 in. WT
Pipe coating — negative buoyancy . . . . to 65 lbs./ft.

Regulatory Compliance . . . . American Bureau of Shipping+ A1 Barge

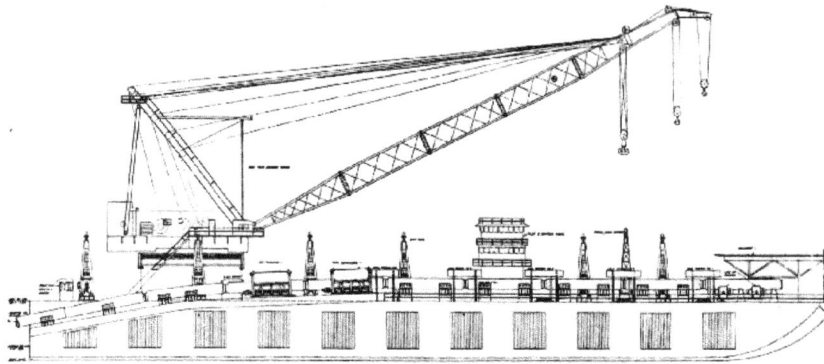

Outboard Profile

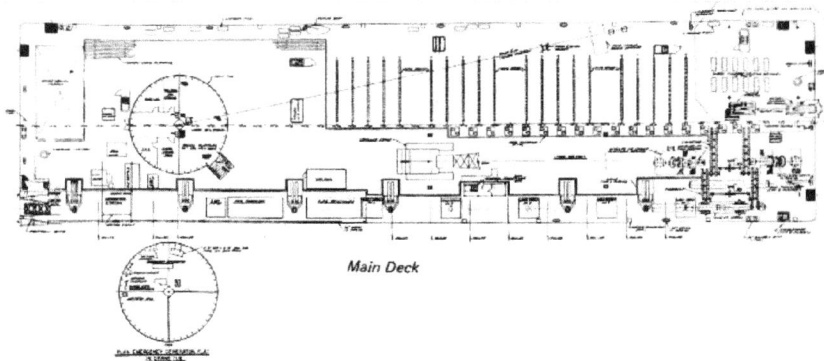

Main Deck

Diagram 7:
ETA's design proposal for a pipelay/derrick barge
Source: ETA Brochure, 1974

Diagram 6 shows the ETA S-5 design of semisubmersible, which was intended to appeal to the same drilling contractor market we dealt with for jackup business.

Almost simultaneously, ETA created a design for a pipelay/derrick barge by building on previous pipelay stress analyses work. This design is shown in Diagram 7.

Both of these designs were really a basis from which to develop a serious proposal for engineering work. Whether it was market timing or ETA simply did not have enough experience, neither the pipelay barge nor ETA's S-5 semisubmersible drilling unit design attracted serious client interest. Coincidentally, they combined to generate interest in 1975-1977 for a semisubmersible pipelay/derrick barge that was discussed with Nippon Kokan K.K. (NKK).

The general thrust in ETA was shifting. We attempted to build more of a general offshore engineering capability. ETA had a name for innovative jackups, and we believed we had the expertise to tackle work on other types of floating equipment on a fee-paid basis. Our growing name as a jackup designer attracted business from Pool Company, which retained ETA to provide jackup designs.

Worldwide Interest In ETA Jackup Designs

After the success of the design contracts for the ETA Robray 300 Class and ETA Europe Class jackups, many shipyards in the world were intrigued by the notion of being able to employ an independent design. Before they generally had to rely on the Big Four jackup designer-builders located in the U.S. Gulf Coast.

ETA embarked on attempts to capitalize on that interest. The overall strategy was obvious: marketing and educating potential builders about ETA and its jackups and laying the groundwork to show shipyards how they might be able to offer construction of jackups of ETA design. Talking to multiple shipyards and doing direct business with some meant a lot of travel. Sometimes potential builders were fabrication yards that normally built production structures but might be able to tackle the complexities of jackups. One example was the yard established three years earlier by Redpath Dorman Long in Methil, Fife in Scotland, close to where I grew up.

Dyvi jackups would go to work in the Norwegian sector of the North Sea, and it was feasible that a jackup built in a UK yard might have more favorable treatment for employment in the British sector of the North Sea. In February 1975, Redpath Dorman Long (RDL) worked up an offer for any organization looking to enter the offshore drilling business in the UK sector of the North Sea. Essentially, they wanted a duplicate of the business that CFEM had for Dyvi's two jackups. RDL cited a price of £17.8 million for their work on first unit and £15.9 million for the second. They estimated an actual construction time of 13 months for each, dependent on timely delivery of all Owner-Furnished Equipment (OFE). It was a truly competitive offer and a hot possibility for a few months but there were no takers and it went dead. RDL is still in business today, tackling offshore wind- and current-driven power-generating platforms.

Some of our business development was done via teaming with shipbrokers. These shipbrokers were well known to the shipyards and were in the business of bringing owners and builders together. To complement the broker's commercial knowledge, they needed to team with an ETA expert, who provided the technical side. Ralph McTaggart, ETA's chief naval architect, travelled to yards in Northern England, Finland, France, and Greece in 1975-1976 for such discussions, teaming with a broker from Fearnleys. It was not always as simple as it might sound. The worldwide ignorance about MODUs— and jackups in particular— meant basic education on jackups was often necessary.

I did expeditions in 1974-1975 to potential builders in South Africa which in that era was remarkably efficient in just about every aspect of business compared to the disarray in West Africa despite the prospering offshore petroleum, industry in that region. South Africa followed a mostly British business culture during these years. I visited Durban, Johannesburg, and Capetown. On a weekend I visited the Agulhas National Park at the extreme tip of Africa and could see the ships on the horizon as they continuously rounded the southernmost tip of Africa in both directions. And a monkey almost stole my camera!

Meetings were held with potential jackup builders, with a marine company and service vessel owner that might be interested in owning and operating jackups, and with financial institutions to get an early idea of financing packages offered in South Africa to attract foreign business. Nothing ever happened to benefit ETA from these early meetings. We did succeed in educating them on the world of offshore drilling and jackups, but that was about all.

In 1974, I travelled to South Korea. As the only Westerner aboard a Fokker F27 on an internal flight from Seoul to Busan, I was uncomfortable in the cramped small seats, which were fine for the locals. Meetings did not pay off. Cultural and technical challenges caused the business trip to not be successful. It was premature; the groundwork of practical education on jackups and the U.S. needed to be laid.

In Japan, I experienced similar results. I woke up at 4 in the morning to my Tokyo Hilton room swaying silently back and forth from a mild category 4 earthquake. No business resulted from this trip. Travel generated considerable interest but some skepticism toward a new outfit like ETA. The reticence was probably justified, since jackups represented serious financial and operating commitment for a drilling company and they were more difficult to build than ships and other marine equipment.

A missionary marketing program takes time as prospective customers weigh their options. Major capital commitments do not happen immediately. The telex machine back in Houston got pretty busy with questions as we helped educate shipyards. I made better headway in Singapore. Their shipyards still used bamboo scaffolding in 1974. The clouds of black smoke from diesel trucks and buses left one headachy on arrival at the shipyard. Singapore is totally different nowadays, with safe scaffolding and elevators, rigorous safety management of all construction, and clean air.) The ETA Asia Class design shown in Diagram 8 was somewhat smaller than the ETA Robray 300 Class units that that Robin Shipyard was building in Singapore and more intended to compete with the LeTourneau 52 Class jackups. Talks started with Far East Levingston Shipyard Pte. Ltd. (FELS) to build the ETA Asia Class on license. An agreement followed, and Picture 9 shows the FELS marketing piece. However, that license agreement did not lead to any construction contracts.

During these travels, I had built a network of ETA representatives in parts of the world where we felt a local representative might generate business: Oslo, Norway; Rio de Janeiro, Brazil; Cape Town, South Africa; and Edinburgh, UK. They were contacts I had made while visiting potential jackup building yards and potential jackup owners. As 1975 approached, conversations with various shipyards about licensing ETA's jackups continued, but the market was slowing.

Additionally, ETA was busy with the business it already had, and I realized our rapid growth was not sustainable at the current pace. We needed to limit the range of what we tackled and control and prioritize our costs. Educating the market was seductive, but it had to lead to serious business prospects. We were impatient for results in what we did but had failed to fully realize it was a game that took some time and business judgement. We needed maturation of the business and focus to spend time and resources as effectively as possible.

THE ETA ASIA CLASS JACK-UP

PRINCIPAL CHARACTERISTICS

Hull
Length . . . . 187 ft. *(60 m.)*
Beam . . . . : . . . . .158 ft. *(48.1 m.)*
Depth . . . . 24 ft. *(7.3 m.)*

Legs
No. 3
Length . . . . 399 ft. *(121.6 m.)*
Type . . . . . . Independent, open truss, with spud cans

Quarters . . . . 78 persons + hospital

Heliport . . . . For Sikorsky S-61N

Rated Drill Depth . . . . 25,000 ft. *(7,620 m.)*

Drill Slot Area . . . . 40 x 52 ft. *(12.2 x 15.8 m.)*

Regulatory Compliance . . . . American Bureau of
Shipping+ A1 Self-elevating Drilling Unit

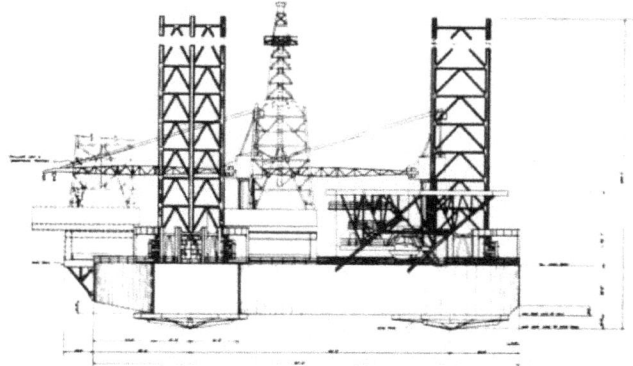

Outboard Profile

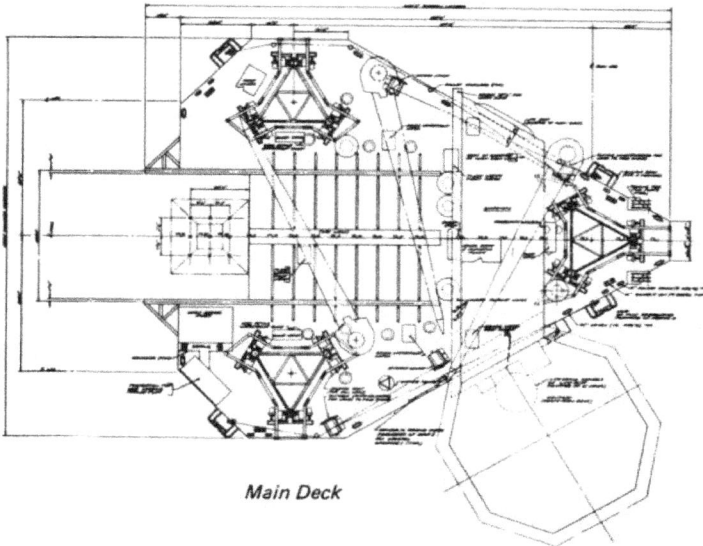

Main Deck

DESIGN CRITERIA

Water depth — ft.*(m.)*	250 *(76)*	275 *(83.3)*
Wave height — ft.*(m.)*	50 *(15.2)*	35 *(10.7)*
Wave period — sec.	12	10
Max. sustained wind — mph*(m./sec.)*	110 *(49)*	110 *(49)*
Air gap — ft.*(m.)*	40 *(12.2)*	30 *(9.1)*
Penetration — ft.*(m.)*	30 *(9.1)*	40 *(12.2)*
Overall leg length — ft.*(m.)*	368 *(112.1)*	399 *(121.6)*

ETA Engineers, Inc.
ETA Building
6101 Southwest Freeway
Houston, Texas 77057 U.S.A.

Telephone: (713) 667-6100
TWX: 910-881-6285
Cable: ETAHOUSTON

Diagram 8:
ETA Asia Class jackup design
Source: ETA brochure

<u>Picture 9</u>:
Image used by FELS to show their ability to build the ETA Asia Class jackup
Source: <u>FELS</u>

ETA Offshore Seminars

One of the things we had embarked on was a seminar venture–kind of fun and good PR, but I would often wonder what exactly it did to translate that "feel good" into billable business.

Some of my colleagues at ETA had an idea to encourage new business by getting ETA better known, building trust, and widening our range of clients. They wanted to hold workshops and network. During late 1972, ETA had formed an affiliate called ETA Offshore Seminars, Inc. The idea was that staff and business friends would speak about ETA and the offshore industry in both Houston and locations in the Far East and Europe where interest was growing in offshore drilling.

It stumbled a bit in the first year because the initial manager came from outside the industry. Then we brought in Ralph McTaggart, our chief naval architect, to lead the effort. That change was what was needed; we didn't need some marketing type. We needed someone steeped in the business, with the people skills that instinctively understood clients' interests.

The seminars were a platform with excellent exposure for the speakers in a fast growing business, and the events themselves turned out to be fun—many speakers enjoyed sharing their knowledge rather than being cooped in an office.

In an era when many people were unaware of what happened on offshore drilling rigs, the seminars were a valuable self-supporting marketing ploy for ETA. One Singapore shipyard manager who attended one of the seminars, said it was "all Greek to me, black magic."

We published a book with materials from these seminars: "The Technology of Offshore Drilling, Completion and Production," Pennwell Publishing, 1976, 426 pages. The dust cover is shown in Picture 10. Contributors in the book included many industry authorities, some still known and around today, such as Mark Childers, Hugh Elkins, Mike Hughes, and even Peter Lovie! I did my part with a rather boring chapter about securing design approvals from classification societies and governmental regulatory organizations.

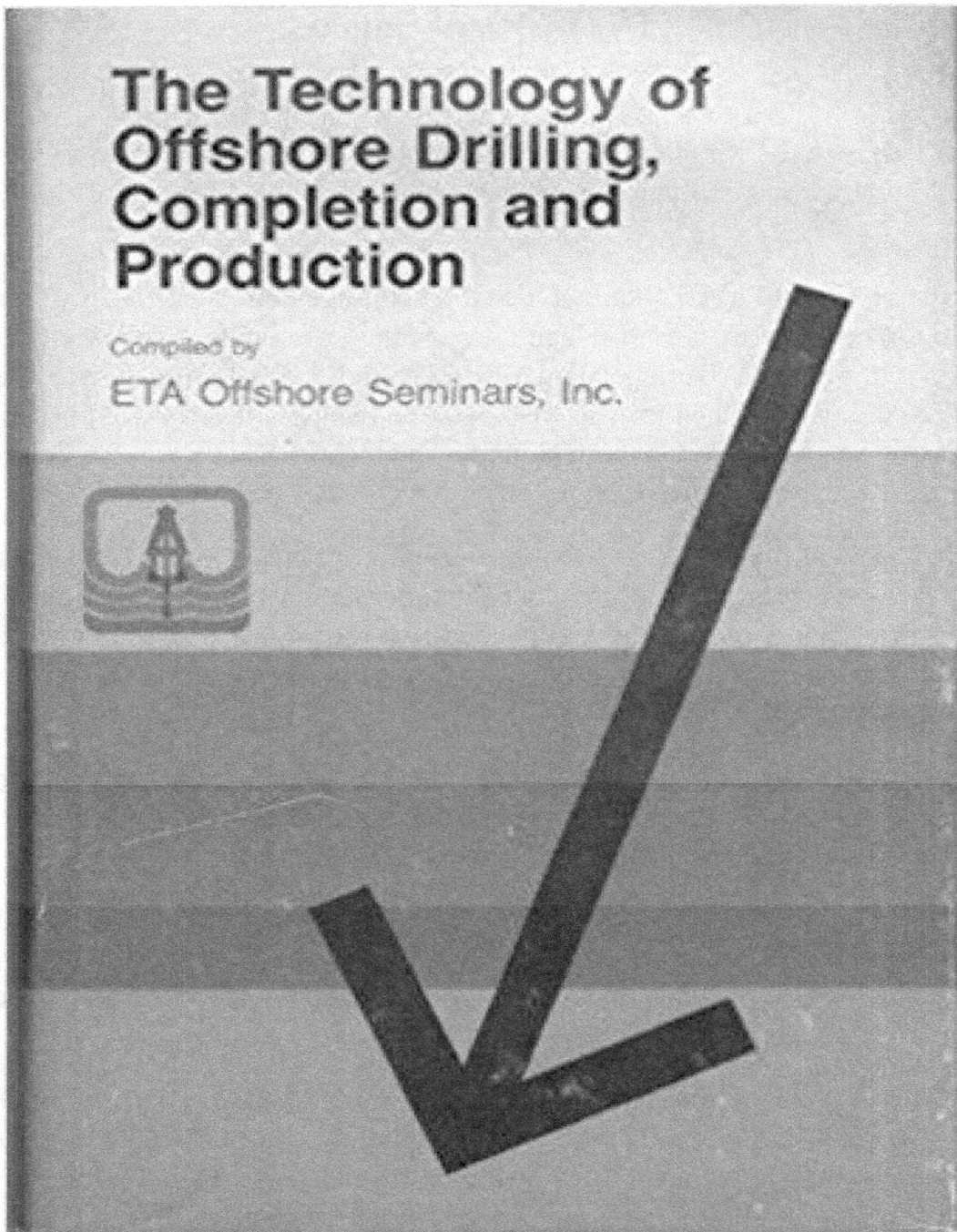

Picture 10:
ETA Offshore Seminars
Source: The Technology of Offshore Drilling, Completion and Production,
Pennwell Publishing, 1976, 426 pages

The book was widely used for a decade or more.

Despite the fun of the seminars, there was no escaping that business was getting more difficult to find as we entered 1975.

Competition

When ETA started, our competition was mostly individual consultants and the advisors that computer service providers would employ to help engineers use the standard programs for structural analyses on the service bureaus. These analyses could run up lots of time on computers and were good revenue generators for computer service bureaus! So there were people like Bill Baker at Control Data who advised on using programs and debugging data.

During the first two years, we found a competitor on the same side of town that provided similar series of structural mechanics solutions—Synercom Technology. In 1969, three professionals left McDonnell Automation to form the company and simplify the use of the IBM FRAN structural analysis program. The founders believed there was a market for that simplification and accompanying consulting services.

In that respect, they were different from ETA, which increasingly saw the potential in doing engineering more effectively rather than advancing software usage for structural analyses. We were head-to-head competitors for a year or two, but they veered off into systems for recording utilities locations, which was an emerging market then. About that time, ETA moved more into jackup design and analytical work, so competitive pressure lessened. (Synercom Technology appears to have gone out of business in the late 1970s.)

Our competition became more with marine consulting companies such as Breit Engineering, Korkut Engineers and Friede & Goldman, all in New Orleans. They had been started and staffed by individuals who already had considerable experience in offshore design and marine operations. Their company growth had been slower and much more moderate than that of ETA.

In contrast, ETA had been a something of a cultural and marketing phenomenon. We were a bunch of relatively inexperienced but bright engineers, rather than a more traditional, older staff. We were young and sometimes unaware of the risks of going too far too fast, a serious weakness.

The closest to a direct competitor on jackup design came in late 1974, when Navire of Finland started Scandril Offshore in Houston. Navire was a company which had been building semisubmersible offshore drilling rigs and saw ETA's success with jackups as a business opportunity. Navire wanted to broaden their MODU building business, considering how ETA had attracted the attention of their Scandinavian rig-owning neighbors.

Scandril started conventionally by hiring known and experienced engineers for a team of six fabrication and design people. One came from IHC, a Dutch-based company with a U.S. startup which had been involved in the design and construction of the *Transworld 64* jackup in Ingleside. The *Transworld 64* had been originally designed and contracted with Transworld Drilling, but was ultimately completed by Baker Marine. Another of Scandril's key engineers came from Marathon LeTourneau.

Scandril's publications were a contrast to ETA's: their "brochure" came with a set of resumes and treatises on jackup design and leg fabrication. They had the makings of a sound technical firm but did not seem to have much of a clue about marketing, communicating with customers and bringing in new business.

They were what a large company in Finland might think appropriate for getting into capital projects. Scandril was not entrepreneurial or adaptive to the Houston market. They went out of business in 1978, even though the MODU business was thriving and growing rapidly at that time.

Marine Engineering Systems was another marine engineering company that offered consulting services. From time to time, they would offer to perform jackup design for clients such as Pool Company. Marine Engineering Systems was acquired by Pace Companies in the mid-1970s.

One of the conventional consulting engineering companies serving the production facilities market was James W. Brown & Associates. In 1976, they marketed a jackup design. It was a non-buoyant truss hull with three legs, which would be delivered by barge and jacked up as a drilling and production structure. Unused leg lengths could be cut off. It was a novel idea and subject to a patent application. It did not seem to go anywhere, although others did experiment with advancing a jackup truss hull for short-term production.

During ETA's existence in 1970-1977, there really was no other consulting company offering jackup designs. The closest was Frieda & Goldman in New Orleans, which was successful with semisubmersible designs but did not do jackup design contracts until 1978 onward, with great success.

Design Rights

Our agreement with Robray Offshore Drilling Co. was ETA's first design deal; we had no precedent and no industry templates for rights on future designs. We approached the design assignment as the development of a design from a concept. The design would be owned 80:20 Robray:ETA, with Robray having the right to buy out ETA for $20,000. The ETA Robray 300 Class design was sound, attracted multiple buyers, and performed well. ETA ended up delivering good value. But at its early stage, ETA was not in a position to bargain for good terms.

After the design work was complete, we learned that Robray contracted with Hitachi for two jackups to be built in Japan, and Robray's shipyard affiliate Robin Shipyard in Singapore had also contracted to build additional jackups for third parties. History shows an overall total of nine were actually delivered during 1976-1982.

ETA did not receive income for successive jackups of this design that Robray's affiliate Robin Shipyard built. Ultimately, ETA went out of business in 1977, and Robin Shipyard continued to deliver ETA Robray 300 Class jackups into 1982 (see Part V for a list.)

ETA's agreement with Dyvi Drilling A/S was more of a simple design license agreement. ETA got paid for delivering the design and getting it approved by the vetting parties. It was more favorable for ETA than the agreement with Robray. ETA was able to license that design for future jackup construction. No further copies of the ETA Europe Class jackup were ever built, although Promet of Singapore delivered a cantilever variant, the *Zapata Scotian*, in 1981, with shorter legs (360 ft.) for service offshore Eastern Canada.

CFEM, the builder of the two jackups for Dyvi Drilling, went on to develop its own design of North Sea jackup using a different hull concept. CFEM succeeded in delivering nine of them during 1978-1986 before going out of business.

The third group of jackups that ETA designed were much smaller. ETA developed these jackups for Pool Company in Houston under their operating requirements. In the design service agreement, ETA did the jackup design work, and Pool chose the layout of drilling and related systems onboard the jackup platform. The designs became the property of Pool Company. In turn, Pool contracted for its jackups to be built in different parts of the world. Ten were ultimately delivered during 1976-1982.

There were two designs for Pool Company: the Pool 140 series, followed by the Pool 50 series. Work on the 140 series started in late 1974 and continued into 1975. We developed a three truss leg jackup with a cantilever for about 150 ft. of water for Middle East service.

I had gotten to know the Danish Trade Consul in Houston (Johan Petersen) and introduced him to jackup construction and the potential for a shipyard in his country. That led to a construction contract with Pool Company to build their *Pool Rig 142* at Par Dan in Denmark, a first for Denmark and in a not-so-typical rig building yard. Part V "From Paper To "Iron": Construction & History Of Jackups Built From ETA's Designs" provides more information on the Pool 140 series and a list of them.

The later Pool 50 series design was smaller. It had four tubular legs and a cantilever configuration with a mat. This design was intended for U.S. Gulf of Mexico service. Part V includes more information.

From the beginning we had always pursued business with the established U.S. drilling contractors but they were hesitant to go first with new designs from ETA and so the business with Pool represented an important start in developing business with local drillers.

A Tipping Point (1975)

With engineering work on the ETA Europe Class winding down, ETA's business with Pool Company was valuable to fill the gap. Some of our most recent and more senior hires worked on this business: Ted Barker (ex-Gray Tool) and Derek Scovell (ex-Bethlehem Steel). We had strengthened our design-drafting department with the addition of Mike Luker and Larry Sparks.

The growth of ETA had been entirely self-funded. This is not unusual in engineering practices but was somewhat difficult to manage given ETA's rapid growth and aggressive marketing program worldwide.

My partner and I had contributed lots of sweat but no dollars other than leaving profits to fund growth quarter to quarter and managing them very carefully. There was something sensible about having limited resources to spend on speculative design work and marketing programs. The discipline of these constraints had worked effectively as we grew.

Industry attention in 1974 encouraged ETA to attempt additional lines of business. We undertook analyses of pipelaying operations. We did not have fundamental experience with semisubmersible design but attempted to apply the talents of one of our naval architects in the design of a compact semisubmersible design that in effect was a competitor to the Breit design.

We had not realized just how much the MODU business fluctuated and how requirements for studies and new designs could easily dry up for several months. We had not thought out longer-term contingency plans to weather downturns. Our naive belief was that we could manage our way through all adversities! And somehow we would find a way to pay the bills, as we always had in the last four years.

Nevertheless, the Pool Company business was a sign that we were being accepted as a viable design engineering outfit, and we attempted to find more of that relatively routine but steadier business.

ETA had grown in employees and needed more space, so we moved a third time at the end of 1974 to 6101 Southwest Freeway. We occupied enough of the building that we could name it the "ETA Building." We thought we were big shots!

As 1975 progressed, it became clear that we were biting off more than we could chew. We struggled with adjusting and reorganizing to be more of a regular engineering services firm. We started to see our success slow.

Chasing around the world in an exciting yet exhausting pace had meant I'd not paid enough attention to ongoing business in Houston that had become increasingly managed by my partner.

My partner felt we could obtain the business to stay alive by hiring new professional managers that did not necessarily have direct experience. I felt we needed to stick with who we had, and save on costs and we could not afford more management overhead.

I felt he was loading ETA costs unfairly from his various other ventures while he saw his own separate business interests were not prospering like ETA.

There was a basic clash of business styles and interests. Conflicts grew. In mid-1975, we were forced to retrench in a down market by laying people off and reversing growth. Parting company with well-qualified colleagues was difficult but something that had to be done for the sake of the business.

Peter Lovie Leaves ETA

The serious difficulties continued with my business partner on spending, priorities, company directions, and inflated costs of his affiliated interests. It was a matter of direction and control and stopping and turning around monthly operating losses. My goals had been treating people right, inspiring them to accomplish big things, and expecting a little fairness back, and it had served me well at ETA. Professional engineers acting professionally could make a business success in an engineering firm–or so I felt. However, despite a promising future, I was embroiled in an immediate downward spiral. I could not see a way of turning it around.

It was three years before Kenny Rogers wrote "The Gambler" with the refrain "Know when to hold 'em, know when to fold 'em, know when to walk away and know when to run." I took some legal steps to limit my financial downside exposure, and decided to bail out, i.e. "to fold 'em and to walk away." It is easy to describe it that way today but weighing it up and applying Kenny Roger's business principles was not so easy back in 1975 after devoting so much energy to it all in the prior six years.

I left ETA in late 1975 with a good name in the industry, great experience, but not a whole lot else. There simply was not much left financially in ETA by that time, and the potential for real financial problems loomed. Despite my warnings, my partner felt he was smart enough to turn it all around!

After always living in apartments since arriving in Houston, in late 1974 I had moved into a much larger place – a house with a deck, with many trees that looked out over the jungle and ravine that was Buffalo Bayou, giving a sense of being away from it all while still in the big city. Close in and off Memorial, neighbors turned out to be mostly well established Houstonians. At the time it just felt like the right thing to do although looking back today it was a modest sign of making it in America.

A room was set aside as an office with desk and filing cabinets. A landline phone was installed and a teletype put in to create a base from which Lovie & Co. did business development for the next four years, arranging many jackup construction contracts, each worth many millions. It was a time when secretarial services and copying companies existed to do letters and documents to support small businesses like mine, decades before Microsoft Word and PCs and doing more on one's own. And at night I had to close the door to the corporate headquarters to stifle the noisy clatter of telexes arriving from different time zones around the world.

PART II

ETA COLLAPSES

Renamed ETA Engineers Inc. in 1975, ETA published a brochure on the ETA Eager Beaver jackup, a design for up to 180 ft. of water, a mat unit with three legs that was intended for drilling in moderate environments such as the U.S. Gulf of Mexico (shown in Diagram 9).

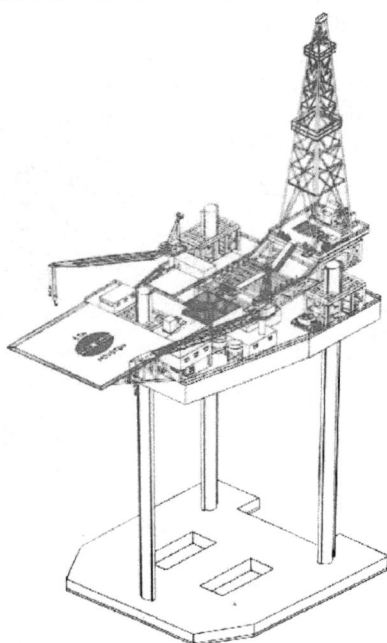

DESIGN CRITERIA

Rated Drilling Depth	20,000 ft.
Environmental Criteria (applied simultaneously)	
Maximum water depth (including tides)	180 ft.
Minimum water depth (including tides)	10 ft.
Design wave height	40
Design wave period	14
Design wave current	
Design wind velocity (one minute sustained)	100 mph.
Design temperature	
Structure	0° F minimum
Quarters	0° to 120° F
Regulatory Compliance	ABS + A-1
Air Gap	30 ft.
Mat Penetration (excluding skirt)	2 ft.
Towing Characteristics	
No leg removal	
Towing draft (upper hull)	10 ft.
Vessel can float on mat in shallow water areas.	

STORAGE CAPACITIES*

Drill water	2,212 bbl.
Potable water	400 bbl.
Fuel oil	1,106 bbl.
Liquid mud	1,942 bbl.
Bulk storage	7,200 cu. ft
Sack storage	1,500 cu. ft
Drill pipe	12,000 ft.

*Maximum consumables capacity, 2,000 kips

GENERAL DIMENSIONS

Upper Hull	
Length overall	150 ft. 0 in.
Beam overall	120 ft. 0 in.
Depth	16 ft. 0 in.
Mat	
Length overall	170 ft. 0 in.
Beam overall	160 ft. 0 in.
Depth (including skirt)	10 ft. 0 in.
Skirt	2 ft. 0 in.
Legs	
Number	3
Type	Caisson
Outside diameter	9 ft. 4 in.
Length	255 ft. 0 in.
Heliport	
Sized to accommodate Sikorsky S-61N	70 ft. 0 in. × 70 ft. 0 in.
Quarters Capacity (includes hospital)	56 persons

EQUIPMENT SUMMARY

Power
Engines: 5 @ 1,325 hp with 800 KW 600 VAC generators
SCR power conversion modules: 4
Emergency generator: 1 @ 250 KW

Drilling
Drawworks: 1 @ 2,000 hp for 20,000 ft. drilling depth
Pumps: 2 Triplex slush pumps @ 1,500 hp each
Derrick: 1 dynamic derrick, 30 ft. × 30 ft. × 140 ft., 1,300,000 lb. hook load capacity
Rotary: 1 @ 37-1/2 in. with 700 hp electric drive
BOP: 1 @ 13-5/8 in., 5,000 psi stack with diverter (2 pipe, 1 blind, 1 bag type)

Diagram 9:
Arrangement, principal characteristics of ETA *Eager Beaver* jackup design
Source: ETA Brochure, *July 1975*

It was a similar configuration to the widely accepted Bethlehem mat units with three tubular legs and a mat for drilling service in water depths up to about 225 ft. The ETA Eager Beaver employed a rack and pinion jacking system in contrast to Bethlehem's hydraulic piston system. It was relatively novel at that time in that it had a cantilever instead of a slot drilling arrangement, so it could be used for production drilling, well service, and traditional exploratory drilling. ETA had chosen to create a smaller, less advanced design than its earlier designs. It was simple and easy to design, but there was no clear, identifiable market need for it that was not already being filled.

There was no talk of where such a design might be built. The intent appeared to be to design something with a minimum of new technology— as the new general manager at ETA liked to say, "we stay away from pushing back the foreskin of science."

It also meant that the engineering was much easier to supervise, was simpler and far less of the kind of challenge that had inspired the "brainy millennials" earlier and who were now leaving the company.

I did not recall there being any basic market rationale behind it, of the kind we had seen in 1972 in entering the jackup design arena. It was a departure in philosophy from the successful precedent of deepwater jackups exemplified by the ETA Robray 300 and ETA Europe Classes of design, which had stretched industry limits and responded to a previously identified market need.

In 1976-1977, I was away from ETA and busy developing jackup construction business for two shipyards, so I quickly learned of efforts by other independent jackup design firms to enter the same market sector that ETA had pioneered, i.e. jackups for deep and often harsh environments. Marine Structures Consultants (MSC) in the Netherlands and Friede & Goldman (F&G) in New Orleans were both promoting their deepwater jackups with their series of CJ designs and L780 design series respectively. Just a year or two later, they both started securing business for their designs. ETA had been on the right track!

The market showed ETA was now going in the wrong direction with its Eager Beaver and spending scarce resources to do so. ETA's general manager liked to affect a Texas "good ol' boy" style in what he said—one expression commonly used for really stupid things or people was that they were "et up with the dumb ass" and that came in my view to describe ETA's actions more and more during 1976-1977!

Frustrating though it might be to see a company I'd helped build now blunder its way downward, I kept distant and quiet. The business with Pool Company had been valuable in broadening ETA's experience with well service and workover business. Engineering the Pool 142 and Pool 50 jackups provided experience in designing cantilever and well service jackups. These two designs were applied in eight more Pool Company jackups delivered in 1978-1982. Pool was the kind of client that ETA needed because of its potential for a long-term service relationship.

After completing work for Pool 50, ETA decided to develop a smaller variant of the ETA Eager Beaver called the Eager Beaver 120 Class, which was intended to work in waters up to 120 ft.. It had four tubular legs instead of three and would serve a jackup market sector similar to the Pool 50, which worked at a 90-ft. maximum water depth. Like the Pool 50, ETA settled on a design with a mat, and each leg had a similar arrangement of rack and pinion elevating system. In 1976, the design appeared in an ETA brochure as the ETA Beaver 120 (shown in Diagram 10).

By normal consulting and marketing standards, investing in the creation of the Eager Beaver 120 Class design was questionable judgment, but matters took a turn for the worse as it jeopardized the relationship with a good client.

Pool Company did not take kindly to seeing the design of the ETA Beaver 120 Class jackup being marketed to its competitors. Pool therefore sued ETA to stop promoting that design, claiming that ETA was copying their proprietary design which they had paid ETA to develop.

It went to trial, and ETA incurred serious expenditures of management time and legal fees to defend their position. The conflict culminated in several days in a court hearing in Dallas, crippling ETA's survival. However, ETA won the case and was awarded a settlement. Collecting turned out to be difficult because Pool promptly countersued.

If you knew the egos and combative natures of Frank Pool and Ed Lowery, quite apart from the merits of the case, it became a tough conflict to settle!

With the benefit hindsight, it was amazing how a design engineering company like ETA attempted to promote such a similar design without at least talking with Pool to avoid overlaps and possible conflicts of interest. The engineering was relatively simple and would have meant far less revenue for ETA than previous deepwater. It was potentially good business for ETA but was not worth risking a big fight!

No ETA Beaver 120 Class jackups were ever built – or any Eager Beaver jackups either.

Beavers did indeed turn out to be toxic critters for ETA, contributing to bringing down the company. The "Eager Beaver" name chosen for its "good ol' boy" double meaning was not funny anymore.

THE ETA BEAVER–120 CLASS JACK-UP

PRINCIPAL CHARACTERISTICS

Hull
Length . . . . 131 ft. *(40 m.)*
Beam . . . . 105 ft. *(32 m.)*
Depth . . . . 12 ft. *(3.6 m.)*

Legs
No. 4
Length . . . . .160 ft. *(48.8 m.)*
Type . . . . Caisson with mat support*
or
Independent truss with spud cans

Mat*
Length . . . . .155 ft. *(47.2 m.)*
Beam . . . . .120 ft. *(36.5 m.)*
Depth . . . . 6 ft. *(1.8 m.)*
Skirt . . . . 1.5 ft. *(0.5 m.)*

Quarters . . . . 35 persons

Heliport . . . . Sized for Sikorsky S-62

Rated Drill Depth .15,000 ft.

Regulatory Compliance A.B.S.+ A1 Self-elevating Drilling Unit

*At the option of the Owner.

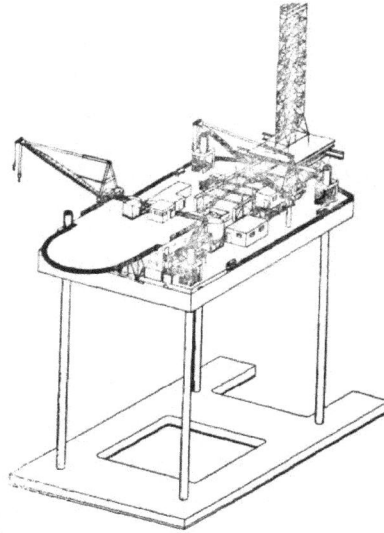

DESIGN CRITERIA

(Applied simultaneously)

Water depth . . . . .120 ft. *(36.5 m.*
Wave height . . . . 25 ft. *(7.6 m.*
Wave period . . . . 7 sec
Wind speed . . . . 70 mph *(31.3 m./sec.*
Current . . . . 1.1 mph *(0.5 m./sec.*
Air gap . . . . 20 ft. *(6.1 m.*

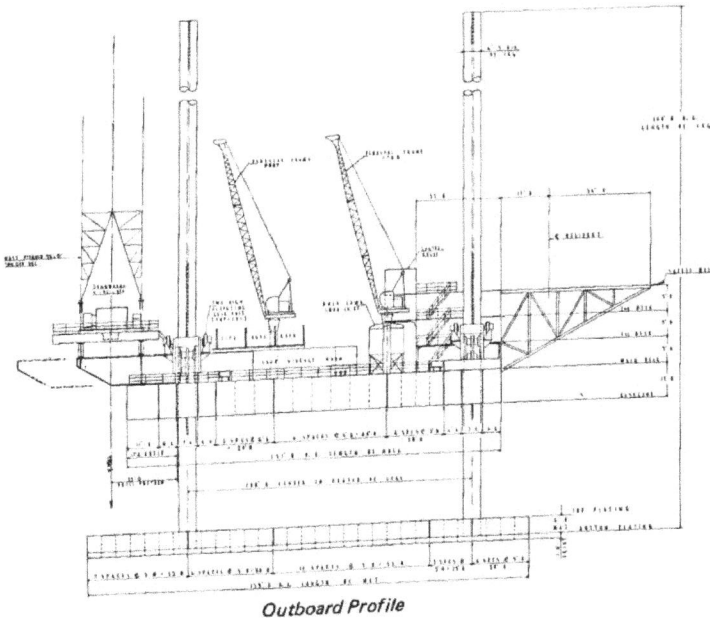

Outboard Profile

<u>Diagram 10</u>;
The ETA Beaver 120 Class, subject of a client's lawsuit
Source: <u>ETA brochure</u>

Downfall To Collapse (1976-1977)

After Hitachi began building the ETA Robray 300 Class of jackups, another large Japanese shipbuilding group, Nippon Kokan K.K. (NKK), contacted ETA in 1975 to investigate how they could build jackups of ETA's design.

In addition, NKK talked about commissioning the design of a new semisubmersible heavy lift barge. It was a type that had just been introduced in the offshore construction business and was potentially a long-lasting, company-building opportunity.

NKK was put off as they got to know ETA. They learned about the departures of many including myself, the lawsuit with Pool, ETA's financial difficulties, and their interest waned. Nevertheless at that time, the offshore market was still an attractive means of offering existing designs to serve the market. NKK and ETA struck a cooperation agreement, which they announced in an April 1977 release. ETA licensed four jackup designs to NKK: ETA Europe Class, ETA American Class, ETA Asia Class and ETA Beaver Class. The small semisubmersible design and the pipelay/derrick barge design were also included.

ETA brochures were printed again with "NKK-ETA" replacing "ETA." In April 1977, *Ocean Industry* ran a full-page advertisement inviting readers to visit NKK at booth 1573 at the Offshore Technology Conference (OTC), where they could see the new Nomad design of semisubmersible crane barge being developed by NKK and ETA.

But ETA was starting to tank severely. About this time, ETA's monthly salary payments to its representative on the Dyvi jackups at CFEM in France were defaulted on once again. This time he faced jail in France for multiple bounced checks. ETA did not fix the situation. With his wife in the hospital, ETA's representative found himself in France without health insurance coverage. Such neglect on an employer's part was not the way to treat anyone, far less a key employee on ETA's biggest most critical project!

Former colleagues said my former partner had been seen dodging visits to ETA's offices by an armed individual. Supposedly this man was from the New Orleans mob, intent on collecting non-ETA debts.

Meeting payroll became a frequent problem as cash flow tightened. Defaulting on office rent led ETA's landlord to lock the company and its employees out of their offices (some lost personal property in their offices). ETA had to find a smaller space. Employees began urgently seeking other employment. ETA was boxed in from its own actions and was running out of recovery options.

My former partner had controlled ETA after I left and had always been ETA's chief financial officer. But he had not forwarded to the IRS tax payments that had been deducted from employees' salaries and held by him in trust. Therefore, he was personally liable for these monies. The accrued employment tax trust fund had grown to a reported six-figure amount, and the IRS increased its efforts to collect.

Rather than reach a payout settlement with the IRS, as is commonly done in such instances, my former partner and his family abruptly left the country for extradition-proof Costa Rica.

A few weeks later, the designs of ETA were sold at a foreclosure sale after ETA's default on a Small Business Administration (SBA) loan.

These are the circumstances behind the collapse of ETA.

Years later, I would get visits from the FBI trying to obtain information about my former business partner, now a fugitive. Just like on TV, the FBI agent would close my office door, flash his badge, and ask his questions, looking for Edwin L. Lowery, but I had long since lost track of him. The FBI made similar visits to former ETA officers who had been there from 1976 to 1977. When the internet arrived, the FBI would include Edwin L. Lowery in its list of wanted fugitives.

I never did uncover why he took that drastic step of leaving the U.S., whether it was solely to avoid payment of these trust fund employment taxes and the SBA loan— or whether there were other hidden, perhaps nefarious reasons— I don't know. The only thing I found that shed a little light on Ed Lowery's subsequent history was a May 2006 article in Costa Rica's *Tico Times*. A paragraph under "No Lack of Scandals" read as follows:

> *Tico Times writer Ronald Bailey unraveled the complex web of companies operated by U.S. financial consultant Edwin Lowery who enticed investors with unregistered mutual funds and farms through his Investment Shop in the '80s. He was later sued for some $4 million by disgruntled investors and slipped out of Costa Rica in 2002.*

Disposition Of The ETA Designs And Software

When ETA's design and software assets became available in late 1977 in a sale of assets, they were bought by Baker Marine Corporation and used in the creation of Baker Marine Engineers (BME), which would provide engineering services and develop jackup designs for construction or licensing by Baker Marine.

BME took over employment of a few of the remaining ETA staff members and installed Peter Nimmo (ex Bethlehem Steel in Beaumont) as president.

The BME brochure replaced "ETA" with "BME" in the names of ETA-developed software and designs. It also reflected Baker Marine's priorities towards building larger quantities of simpler jackups for moderate or mild environments, not burdened by the fundamentally demanding engineering required for large harsh environment jackups like the *Dyvi Beta* and *Dyvi Gamma,* and all the stringent demands of regulators and classification societies that entailed.

Just months after setting up the cooperation agreement with ETA, NKK replaced "ETA" with "BME" in their brochures. This was an advantageous move for NKK, which could now be linked to a jackup builder while still having access to the ETA designs.

The Baker 150 and 250 Classes of jackup design that were contracted for construction in 1978 onwards shared similarities with ETA's Eager Beaver.

The ETA Europe Class design, which had been ETA's major achievement was relegated in the BME brochure to a small diagram, signaling the end of active marketing for the future use and development of the ETA Europe Class design.

A drilling requirement offshore Eastern Canada led Zapata to contract with Promet in Singapore for construction of a harsh environment jackup able to drill in that region. It resulted in a 1981 delivery of a variant of the ETA Europe Class design which used a cantilever and much shorter legs. *Zapata Scotian* was the only other jackup employing the ETA Europe Class design. Today it is laid up in UK waters at Harwich as the *Paragon 391.*

The year after ETA's collapse, Fried & Goldman (F&G) closed the first contract for one of their L780 designs of 300-ft. water depth jackups. It confirmed there was a place in the market for an independent design firm to do what ETA had been pioneering. F&G went on to do design deals for many jackup designs (more on this in Part VI). F&G

performed very well in their jackup design campaign; Jerry Goldman deserves great credit in steering it skillfully and establishing what became a standard in the jackup industry.

Peter Lovie After ETA

In late 1975, immediately after leaving ETA, I made agreements for marketing the construction of jackups as a U.S. commission sales agent for FELS of Singapore and for BMC of Texas. At the start of 1976, I kicked off this new business, going after the complete construction instead of design and engineering only. It was as high-risk a proposition as before but with the prospect of higher returns at lower overhead.

In this new endeavor, I put my engineering, sales, and marketing experience to work. I built on experience and industry relationships from my ETA years. There were initially dry spells for months, but it went well and paid well— much better than ETA. When there's no business, the shipyards are anxious for help but when you get successful at it and reel in good commissions, they somehow get reluctant to pay. It was a high return, high risk business if ever there was one!

I uncovered in ETA what was a potentially difficult and missing part of jackup business: the concept of a "kit" of design license plus leg chords plus jacks. It was a concept I wrote about in a conference paper in 1977, which is cited in the Appendix at the end of this book. The piece was written about two years before that "kit" concept caught on in the jackup industry and become widespread.

From 1982 onwards, it was a lean time developing business for offshore drilling. Oil prices broke downwards, bringing a long period of bad times in the petroleum industry akin to today's major downturn. For a short while, the world's shipyards worked through their backlogs and hoped for new orders, but there was no escaping the difficult MODU market through the low point of 1986 and well into the 1990s.

I read that Ned Simes, the president of Diamond M Drilling (who I used to call on) had retired in 1985 and then started a winery near Austin. I thought he really got it right!

The website for Grape Creek Vineyards explains how Ned and Nel Simes began a twenty-year journey to create a recognized winery in the Texas Hill Country, how they planted Cabernet Sauvignon and Chardonnay, and established the first underground and commercial barrel cellar in central Texas. Ned passed away in 2004, and Heath Vineyards purchased the property and business from Ned's heirs in early 2006.

My guess is there are others in the offshore drilling industry that wish they had Ned Sime's vision and that kind of happy ending!

The mid-1980s marked diversification for service providers in the petroleum industry into other businesses. It was difficult because of the general low activity and tight margins in Houston. I found it was hard to get paid, even when you did secure business. Even with a judgment in one's favor, the payee went bankrupt in more than one occasion.

Most of us had to downsize our careers and lives, often making a living in crap jobs. Symbolic of the depression in the oilfield, in 1986 I sold my Mercedes 450 SEL, a relic from the good times, to a Chinese guy with a bundle of Benjamins. Then I bought a more sensible used Ford Escort for daily transport.

During 1989-1994, I applied my pioneering instincts to develop subsea processing systems, but history has shown the effort was ten to fifteen years premature.

In January 1995, I took a steady job with Bluewater of Holland and based in Houston as their North American representative for their FPSO and SPM business. This position led to a new, interesting, and more stable career in the floating production world. Seven years later, that job led to a new venture with American Shuttle Tankers (which became part of Teekay). Eventually, I enjoyed a stint with Devon Energy, a large independent operating oil company that had substantial offshore interests domestically and internationally. It was fascinating to be addressing all the factors in the offshore field development business-technical, commercial and basic economics. It went fine until Devon divested itself of offshore assets in 2009 to concentrate on onshore, mostly shale-related business.

Along the way, I contributed to what became a twenty-year campaign to introduce FPSOs in the U.S. Gulf of Mexico. The effort which started in 1996 led to the contracting for the first FPSO in US GoM in August 2007 at a 50:50 Devon:Petrobras development. After the industry-wide delays with *Macondo*, it achieved first oil at *Cascade/Chinook* in February of 2012.

I spoke about that floating production saga at Rice University in 2016. During the talk, the computer controlled audio visual system went dead for 10-15 minutes. I continued anyway as it was a topic I'd lived with for many years. Later I joked with the organizers that it was no sweat; it was just William Marsh Rice, the founder of the university, looking down after a hundred or so years, testing whether speakers at his university could still think on their feet without their modern PowerPoint crutches!

That FPSO story recently became a book: "Why Only Two FPSOs in U.S. Gulf of Mexico? The late start and twenty year saga" in which I talk about the technical, regulatory, economic and political aspects. More is at www.FPSOsinGoM.com.

PART III

DESIGN PHILOSOPHIES BEHIND JACKUPS BUILT TO ETA DESIGNS

"Imagination is more important than knowledge. For knowledge is limited, whereas imagination embraces the entire world, stimulating progress, giving birth to evolution."

—Albert Einstein, What Life Means (1929)

In one way, the jackup design philosophy in ETA was pretty simple: how can we devise a structure that does what existing jackups do but much better, survive more difficult loadings, and do it with as efficient a structure as we can?

Less steel weight obviously meant fewer dollars to build, as existing jackups seemed to be overweight. We just went ahead and did it, not stopping to think of how significant it was. We dreamt up many ideas that at the time we thought could be revolutionary and patented a number of them. But as time passed and we thought more, some of them were not too realistic! The concepts that were not built are covered in Part IV.

None of us in the design loop had had a long career building jackups in a shipyard. Our offshore rig fabrication experience was limited, so fabrication choices were what we felt as structural engineers would be simple and logical. Maybe that was a good thing as we pushed limits. We often thought out of the box, although in the early 1970s no one used that expression.

The *AISC Handbook* and *ABS Rules* were the handbooks everyone used as a starting point in design rules to play by. Hull configuration was pretty basic, a matter of accommodating all the equipment and systems needed for drilling operations, which prioritized finding what was important to the drillers. We made the roughly triangular barge structure strong enough to take the storm and ocean tow loads, as well as operational loadings during drilling, while ensuring we satisfied *ABS Rules* on the hull structure.

Published Jackup Design Standards Of The Day

After contributions from leading offshore drilling contractors, MODU designers, and MODU builders, the first edition of *The ABS Rules for Building and Classing Mobile Offshore Drilling Units* came out in 1968, but it was really something of a moving target as we all learned and adapted in this burgeoning business. A revision quickly followed in 1973, and another followed in 1980.

In addition, we all used the American Institute of Steel Construction (AISC) handbook as a design basis for steel design practices that ABS did not cover.

Bethlehem Steel Corporation in Beaumont, Texas had been actively contributing to the industry with occasional technical papers that were good design and information sources. Their paper at Society of Naval Architects & Marine Engineers (SNAME) in 1957 was a 49-page treatise from three of their design and construction authorities. The piece went through many of the calculations and themes needing investigation.

At that point in offshore drilling history, LeTourneau was making headway in the pioneering and marketing of their jackups: *Scorpion* was delivered in 1956 to Zapata Offshore Company and *Vinegaroon* in 1957.

One can guess that Bethlehem in their 1957 paper was intent to establish their credibility in this new jackup business, competing with LeTourneau, rather than being solely motivated to deliver a professional contribution to industry via SNAME!

Apart from the *ABS Rules*, it was a matter of whatever technical paper might be relevant. For example, ETA's paper on floating stability in the 1973 Transactions of SNAME became widely used. The Offshore Technology Conference (OTC) was often the first source we would check; it provided thick, multi-volume documents which were religiously investigated. Technical sessions at OTC were a must. Today these ancient technical papers can be searched online using SPE's Onepetro system, and all the OTC paper copies can be relegated to recycling!

ETA's Approach

The real opportunity for ingenuity came in the design of the legs.

The first *ABS Rules* of 1968 had been relatively new, just five years old when we started on the ETA Robray 300 Class design, and an update came out later in 1973. When it came to creating the ETA Robray 300 Class design, the 425-ft.-long legs were designed to satisfy the ABS MODU Rules of 1973, employing a criterion of 15 degree each side from vertical for ocean tows. We designed the legs so that top sections did not need to be removed for ocean moves— a serious design advance in 1973! Still, rather than blindly following some set of rules, the whole design process was in our minds a matter of real, fundamental engineering that engineers rarely get to do!

The Robray 300 hull at 27 ft. deep was conventional, a foot deeper than usual for added stability afloat. It was a small change that turned out to be a wise one. Drilling equipment and systems were conventional for exploratory service; the National 1620 drawworks at 3,000 h.p. to pull a 1,300-kip hook load, with heavy dual mud pumps typical of the 1970s. It was a system rated for drilling up to the deepest that operators might ever tackle back then: 30,000 ft.! (For reference, one kip is one kilo pound, or 1,000 lb., or one half of a short ton that is 2,000 lb.)

We put significant thought into the leg design. For this large and long-legged requirement, triangular legs were deliberately chosen over a square configuration for fundamental structural efficiency, even though the precedents of the Levingston 111 Class and Marathon LeTourneau 53 Class used square legs for comparable water depths, apparently driven by yard fabrication considerations.

ETA's many structural analyses of the previous 2 ½ years proved to us that our triangular leg configuration was the best way to go. The leg chords were set further apart (about 35 ft.), which again helped structural efficiency. Materials were state-of-the-art: a 5-in.-thick rack in the chords, 100 ksi yield strength, with tubular bracing members that were 65-or 85-ksi yield.

The One Patent Used In ETA's Designs

ETA was granted five U.S. patents for jackup design, but only one was ever used in the jackups built: the ETA Robray 300 Class design and the ETA Europe Class design. It covered the ETA leg chord design with opposed racks, separated by an interior diametral plate. Tubular leg chords were used with cast steel joints between chord and bracing members and also between bracing members at the K brace connections. To improve fatigue and maximize structural efficiency, cast steel connections had the radii and local thickening needed for stress reduction.

The following patent we filed in 1974 was used in many more jackups than we ever expected:

> "Self-Elevating Offshore Drilling Unit Legs," US patent no. 3,967,457, inventor: Peter M. Lovie, filed July 11, 1974, granted July 6, 1976.

It is quite explicit in the leg design details, which are shown in the diagrams in several of the following pages and claimed in the patent text. Part V identifies the numerous times the designs using features in that patent were used during 1976-1986. Part V also describes some of the fabrication methods for using the cast steel nodes. None of the other four patents were ever used. They are described in Part IV, "Brainstorms, Concepts, and Patents: Paper Only! The Jackup with One Leg that Inspired Others."

Design Philosophies

Temperatures in the Far East were moderate, so brittle fracture was less of an issue but required some different material choices in the North Sea designs that we worked on in the following year. Similarly, wind, wave, and current conditions varied from region to region in the world and were much more demanding in the North Sea than in the Far East.

In the early 1970s, the classification societies were just beginning to really establish what was needed for jackups. Thinking out what ETA would do in its designs, it had been a pioneering effort to get it right and be able to justify principles to the authorities at classification societies, who in reality were also finding their way. There was a lot at stake. Jackups were getting bigger, and here was ETA, this new company with new designs, that owners would look to be reliable, safe and insurable. Only in 2017 did I learn from ABS how they had assigned their top two jackup engineers to check out these new ETA designs. Similarly I learned how underwriters and DNV also took extra care in vetting this new breed of jackup design.

The 1970s were a time when a large part of offshore exploration was in waters of 300 ft. or less, so jackups were almost always the exploratory tools, hence the usual practice of carrying maximum-capacity drilling packages. They were usually built with drilling slots, and the use of a cantilever was rare. Slot jackups were the normal requirement. If an offshore driller wanted to use a cantilever configuration to facilitate getting over a platform for workover or production drilling, it meant paying tribute to The Offshore Company (TOC), which held the patent on the concept after they built their *Hustler* jackup. TOC had a remarkable talent for finding potential users of cantilever configurations and sending out their letter, "hustling" about $15K for a license to employ their patent.

Diagram 11(Next Page):
Elevation and plan for the ETA Robray 300 Class jackup design
Source: U.S. Patent *3,967,457*

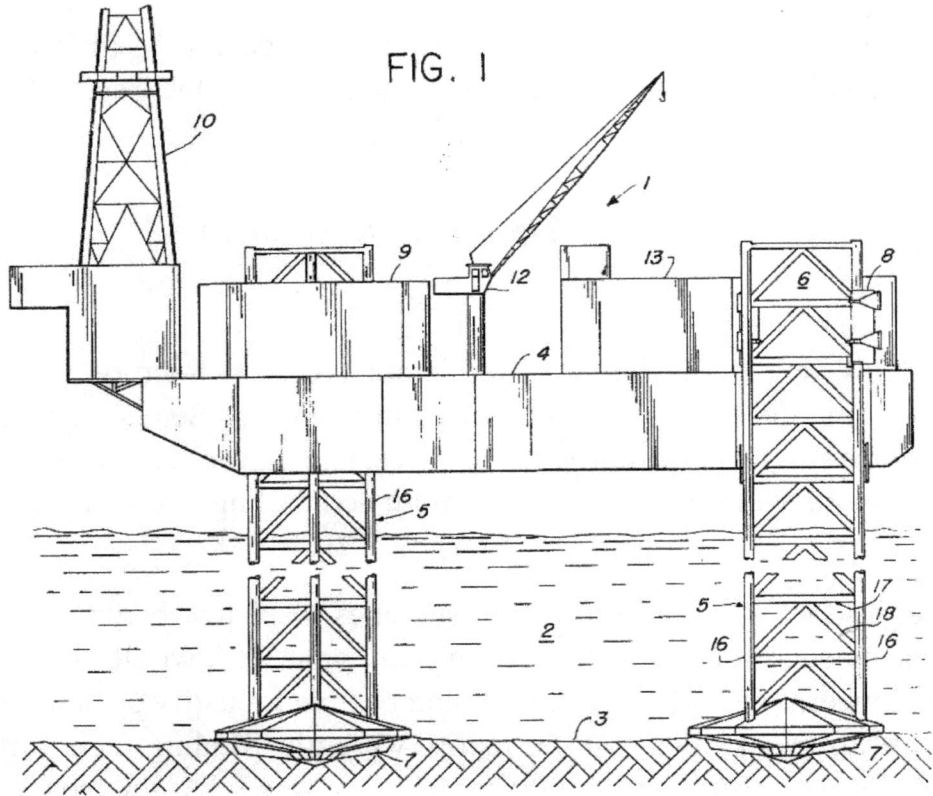

FIG. 1

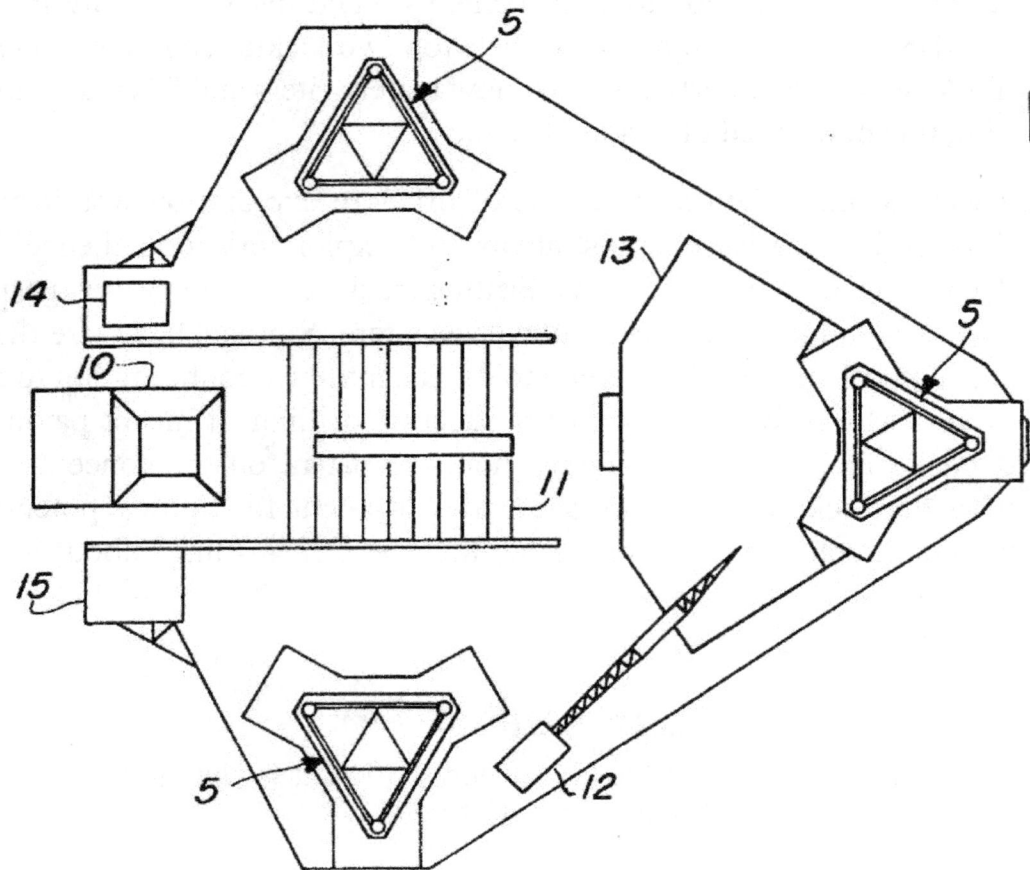

FIG. 2

Forty-four years later, I doubt if any drawings still exist for that original design of the ETA Robray 300 Class, but some idea of its general arrangement can be found from the diagrams in U.S. patent 3,967.457, which applied to the ETA Robray 300 Class design in 1973 and for some features of the ETA Europe Class design in 1974 (shown here in Diagram 11).

As the 1970s moved into the 1980s, more cantilever jackups were contracted for construction. The first wave of jackup deliveries in 1970-1977 was mostly a slot wave, and the second wave of 1978-1986 became a cantilever wave as the versatility of a cantilever jackup became popular to enable both exploratory and production drilling.

Drillers were often gamblers by nature, so you might say they stopped playing the slots around 1977-1978!

The ETA Robray 300 Class and ETA Europe Class arrangements were generally similar except that the aft legs of the ETA Europe Class were turned around so the two chords in the triangular legs faced outwards, increasing leg centers for better overturning resistance (compare Diagrams 11 and 12).

Diagram 12 is a brochure page for the ETA Europe Class jackup showing design criteria and principal characteristics, followed by ETA's original arrangement drawings from 1975 for the ETA Europe Class design used for *Dyvi Beta* and *Dyvi Gamma*. These drawings were thought to be lost but were discovered in a file box in my garage, among a number of boxes of ancient documents that were about to be sent to trash!

Diagram 13 (Page 111):

1975 design drawing for outboard profile of the ETA Europe Class jackup

Source: <u>ETA Drawing</u> *"ETA Europe Class Outboard Profile" Dwg. No. 09630-004, dated January 13, 1975, Revision 2*

Diagram 14 (Page 112):

1975 design drawing for main deck plan of the ETA
Europe Class jackup

Source: <u>ETA Drawing</u> *"ETA Europe Class, Main Deck Plan" Dwg. No. 09630-006, dated January 9, 1975, Revision 2*

PRINCIPAL CHARACTERISTICS

Hull
Length . . . . .230 ft. 3 in. *(70.2 m.)*
Beam . . . . .212 ft. *(64.6 m.)*
Depth . . . . 27 ft. *(8.2 m.)*

Legs
No. 3
Length . . . . .509 ft. *(155 m.)* maximum
Type . . . . Independent, open truss, with spud cans

Quarters . . . . 78 persons + hospital

Heliport . . . . For Sikorsky S-61N

Rated Drill Depth . . . . .20,000 to 30,000 ft. *(6,100 m. to 9,150 m.)*

Drill Slot Area . . . . 50 x 52 ft. *(15.2 x 15.8 m.)*

Regulatory Compliance . . . . . . .Det Norske Veritas+1A1 Self-elevating Unit
Norwegian Maritime Directorate
Department of Energy (U.K.)

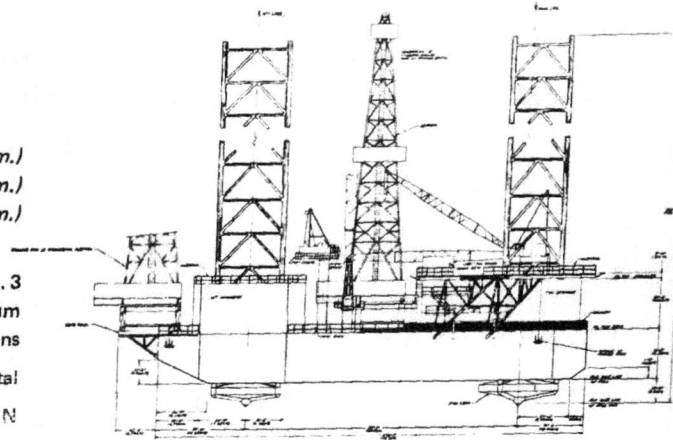

Outboard Profile

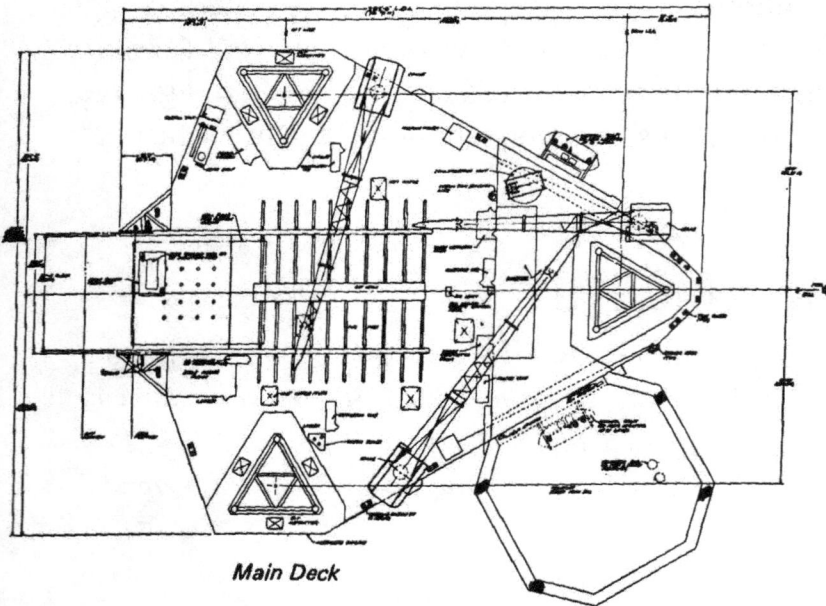

Main Deck

DESIGN CRITERIA

Water depth — ft. *(m.)*	260 *(79)*	300 *(91)*	350 *(106)*
Wave height — ft. *(m.)*	75 *(22.8)*	70 *(21.3)*	65 *(19.8)*
Wave period — sec.	14	13	12
Max. sustained wind — mph *(m./sec.)*	110 *(49)*	95 *(42.5)*	80 *(35.8)*
Air gap — ft.*(m.)*	55 *(16.8)*	55 *(16.8)*	50 *(15.2)*
Penetration — ft.*(m.)*	20 *(6.1)*	50 *(15.2)*	50 *(15.2)*
Overall leg length — ft.*(m.)*	395 *(120.3)*	467 *(142.3)*	509 *(155.1)*

<u>Diagram 12</u>:
Arrangement and principal characteristics of the ETA Europe Class Jackup
Source: <u>ETA brochure</u>

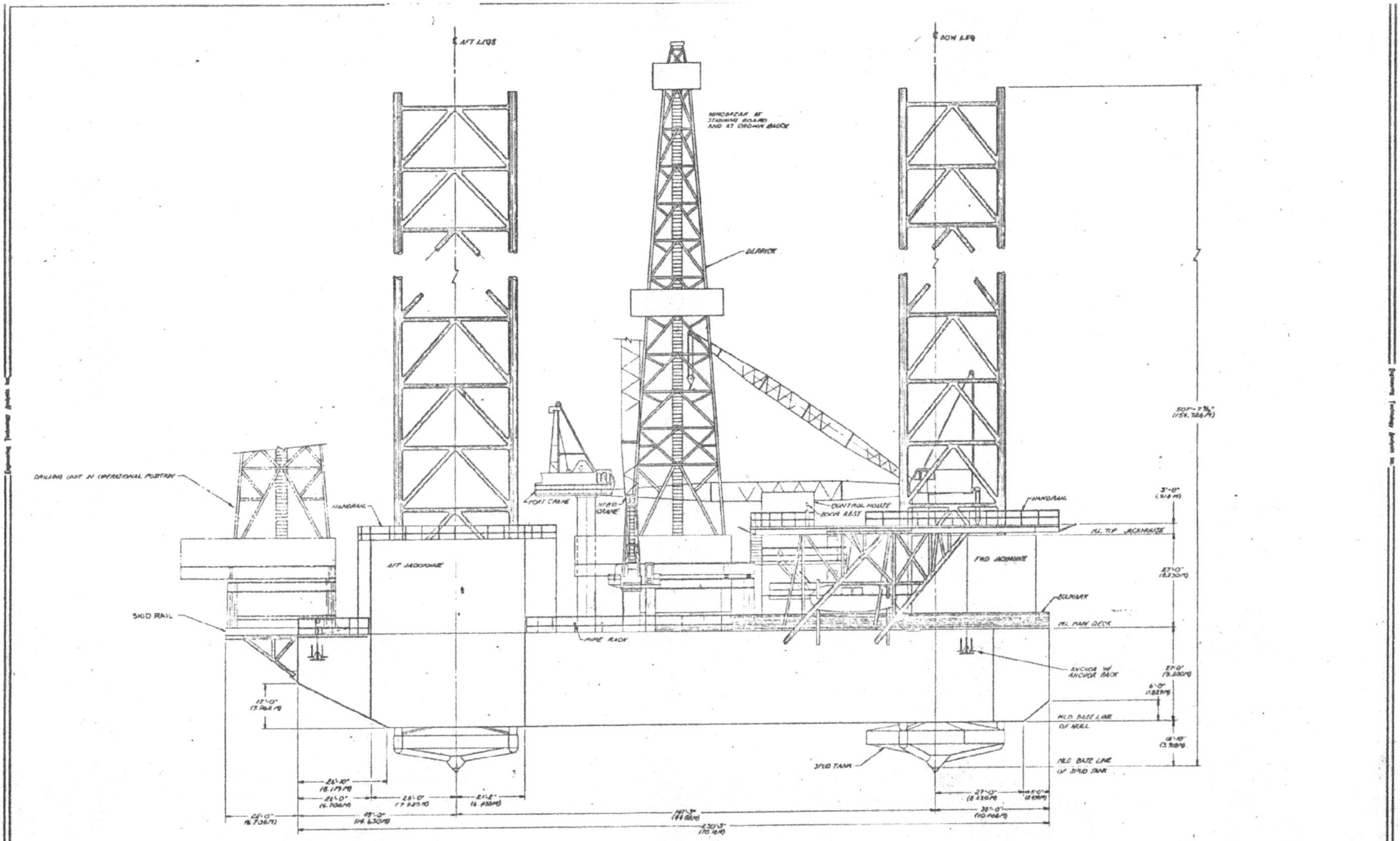

ETA EUROPE CLASS—
—OUTBOARD PROFILE—

Engineering Technology Analysts, Inc.

Engineers

Houston
Texas

DRAWING NUMBER 09630-004

REF 2

REVISION

NO	DATE	DESCRIPTION	BY	CHK'D
1	4-6-74	REVISED TO SUIT DESIGN DEVELOPMENT		JMD
2	1-13-75	REVISED TO SUIT DESIGN DEVELOPMENT	JD	DG

Labels visible on drawing:
- WINDLASS OF STANDING OUTBOARD AND AT CROWN BARGE
- GULLIVER
- CONTROL HOUSE ROOM REST
- UNIVERSAL
- MAIN TOP JACKHOUSE
- FWD JACKHOUSE
- GANGWAY
- MAIN MAIN DECK
- ANCHOR RACK
- ANCHOR DAV
- MLD BASE LINE OF HULL
- MLD BASE LINE OF SPAD TANK
- SPAD TANK
- AFT JACKHOUSE
- CRANE RACK
- SERVICE CRANE
- LIGHT CRANE
- HANDRAIL
- SKID RAIL
- DRILLING UNIT AT OPERATIONAL POSITION

ETA EUROPE CLASS
MAIN DECK PLAN

DRAWING NUMBER 09630-006

Engineering Technology Analysts, Inc.

PROPRIETARY INFORMATION

Diagrams 13 and 14 are both shown landscape-style to make them as large and as readable as possible. If one looks closely or magnifies the drawing onscreen, it is possible to read most of the notes and dimensions.

The two arrangement drawings demonstrate that the ETA Europe Class hull layout was not much different from other three-legged jackups in equipment arrangement, the drilling slot location, and how the drilling package was deployed and the living quarters were located forward. Just as with the general arrangement of the overall jackup designs, the next few pages start with the patent diagrams, which show the principles of the design for the cast steel joints and for the legs. It is illustrated with actual design drawings for the cast steel joints and legs.

ETA's Rack And Pinon Elevating System

The ETA Robray 300 Class jackups employed the National Supply elevating system, which was sold as part of a larger equipment order placed for Owner-Furnished Equipment (OFE) as it was called back then. Robray Offshore Drilling Co. in Singapore arranged a big OFE purchase for multiple rigs, with some of the equipment destined for Robin Shipyard ultimately being sent to markets that were then difficult or prohibited from the U.S., i.e. North Korea and China.

Diagram 15 shows the arrangement of the jacks and the leg chords. This diagram is again extracted from the patent and shows general similarities with the ETA Robray 300 Class and ETA Europe Class designs.

The chord design with the two opposed racks is shown in Diagram 16, separated by the diametral plate. Diagram 16 also shows the K bracing configuration.

Similarities Of ETA Europe Class And ETA Robray 300 Class Designs

The design of the ETA Robray 300 Class and ETA Europe Class are obviously similar in many ways, so key elements are compared in Table D.

United States Patent [19]

Lovie

[11] **3,967,457**

[45] **July 6, 1976**

[54] **SELF-ELEVATING OFFSHORE DRILLING UNIT LEGS**

[75] Inventor: **Peter M. Lovie**, Houston, Tex.

[73] Assignee: **Engineering Technology Analysts, Inc.**, Houston, Tex.

[22] Filed: **July 11, 1974**

[21] Appl. No.: **487,574**

[52] U.S. Cl. **61/46.5**; 61/53; 52/731; 74/29; 254/106

[51] Int. Cl.² **E02B 17/00**; F16H 19/04

[58] **Field of Search** 61/46.5, 46, 53; 52/731, 638, 726, 720; 37/73; 74/29; 254/106

[56] **References Cited**

UNITED STATES PATENTS

2,578,364	12/1951	Maxon, Jr.	52/638 X
3,183,676	5/1965	Le Tourneau	61/46.5
3,372,907	3/1968	Smulders et al.	61/46.5
3,445,129	5/1969	Penote	52/638
3,779,656	12/1973	Guy et al.	52/638

FOREIGN PATENTS OR APPLICATIONS

511,789	6/1952	Belgium	52/726
921,439	1/1947	France	52/731

Primary Examiner—Jacob Shapiro
Attorney, Agent, or Firm—Torres & Berryhill

[57] **ABSTRACT**

Leg apparatus for an offshore drilling unit of the self-elevating type having a floatable hull and a plurality of legs movable from a raised position, in which the legs are supported by the hull in a body of water, to a lowered position, in which the hull is supported by the legs on the floor of the body of water. The legs may comprise a plurality of mutually parallel tubular chord members rigidly interconnected by structural bracing members. The tubular chord members may comprise an elongated tubular body and an elongated plate member, whose longitudinal axis coincides with the axis of the tubular body, rigidly connected to and spanning the interior of the tubular body.

17 Claims, 10 Drawing Figures

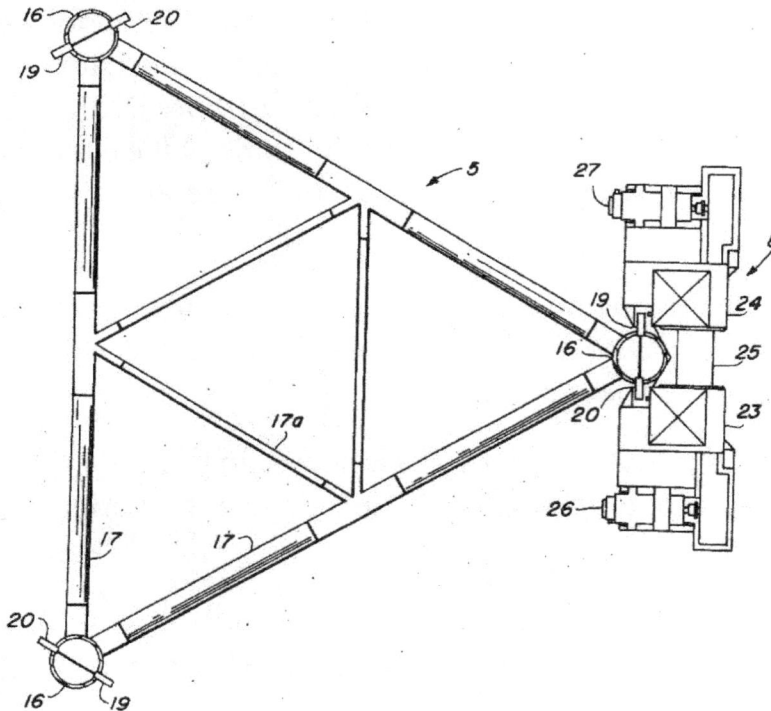

Diagram 15:

Arrangement of ETA jackup leg and rack and pinion elevating system employed at each leg chord

U.S. Patent   July 6, 1976   Sheet 3 of 4   3,967,457

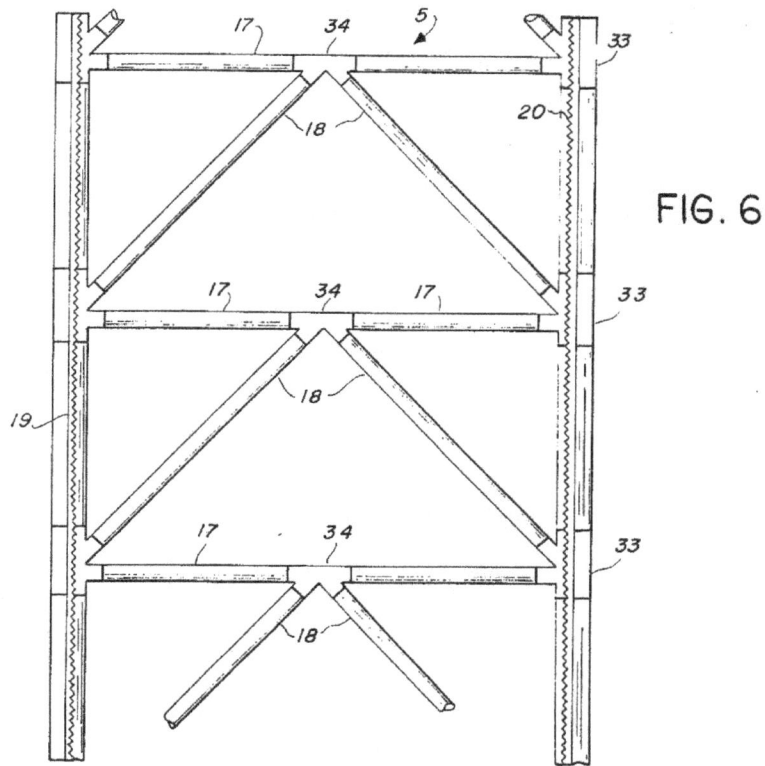

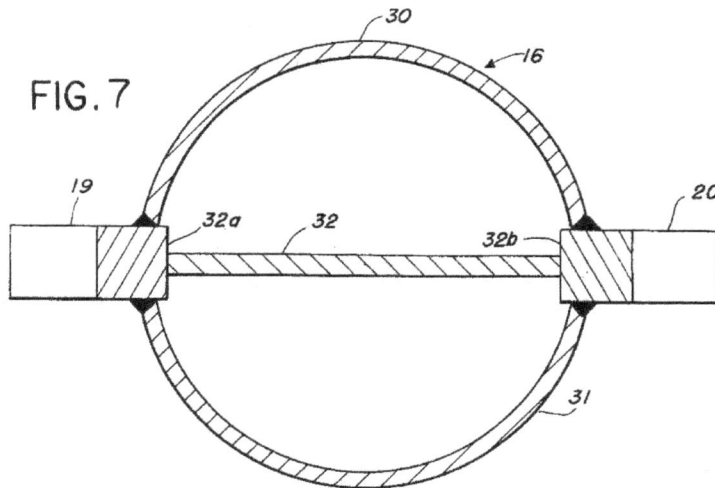

FIG. 6

FIG. 7

Diagram 16:
Typical arrangement of K bracing tubulars in legs, with detail on leg chord design with rack (applies to both ETA Robray 300 Class and ETA Europe Class jackups)

While the drilling systems and the arrangements of legs and hulls had similarities the big differences between the two designs were in leg length (508 v. 425 ft.) and particularly on environmental loads, which were two to three times as severe on the ETA Europe Class; these were at the time believed to be the most severe loads ever tackled. Additionally, the regulatory focus was more stringent in the North Sea than anywhere else.

Later in Part V, all jackups that were built based on each design are listed.

Significance Of The Cast Steel Joints

A key innovation in the leg design for both the ETA Robray 300 Class and the ETA Europe Class designs was the use of cast steel joints in both the connections for bracing members to leg chords and at the K brace connection of horizontals to diagonal bracing members.

Diagram 17 shows the arrangements of these cast steel connections, again using figures extracted from the patent.

At the time, this feature was not thoroughly explained to the industry; therefore, the below explanation is thought to be the first comprehensive written statement of the principles behind it:

1. In the early 1970s, it was known in the industry that there were stress concentrations at the leg joints in truss leg jackups. Strain gage model tests and photoelastic methods for piping joints were known to offer indications of levels of stress concentrations that might occur in pipe intersections of the kind found in jackup truss legs. The stress concentrations were serious, as signaled by the frequency of cracking in jackup legs at the time. Years later, computer methods would improve and finite element analyses became a common way to model connections between leg chords and bracing members, as well as connections between columns and the lower hulls in semisubmersibles, so design details improved. But we had to use the tools available;

Design:	ETA Europe Class	ETA Robray 300 Class
General:		
Year designed	1974	1973
Jackups delivered	2 in 1976-1977	9 in 1976-1982
Builders	CFEM in France	Robin in Singapore, Hitachi in Japan
First Owner	Dyvi Drilling A/S of Norway	Robray Offshore Drilling Co. of Singapore
Status in 2017	See separate table	See separate table
Hull and legs:		
Max. water depth, ft.	350	300
Design leg length, ft.	508	425
Leg centers, fore-aft, ft.	147.3	130
Leg centers, aft, ft.	170	150
Leg configuration	Triangular truss	Triangular truss
Bay height, ft.	18	18
Cast steel leg joints?	used in 2 of 2	used in 4 of 9
Jacking system	rack & pinion	rack & pinion
Jacking speed, ft./min.	1.5	1.6
Hull dimensions, LxBxD ft.	230 x 212 x 27	213 x 212 x 27
Classification	D N V	A B S
Accommodation	83 people	75 people
Variables capacity, kips	9,500	7,400
Drilling systems:		
Hook load, kips	1,300	1,300
Hull configuration	Traditional slot	Traditional slot
Drilling depths, ft.	30,000	30,000
Drawworks, h.p.	3,000	3,000
Mud pumps	2 @ 1,600 hp x 5,000 psi	2 @ 1,600 hp x 5,000 psi

Table D:

Summary comparison of ETA Europe Class and ETA Robray 300 Class designs

2. The inspiration for using cast steel connections came from my first real job in UK in 1965-1966. I had to come up with the same jointing system for all tubular members in a 152-ft.-tall prototype of a geodetic transmission tower structure. Mass-produced castings thus could become reasonable in cost and accommodate changes in wall thickness in tubulars up and down the leg structure;

3. The joint design in LeTourneau and Levingston jackups was apparently driven by fabrication and often had gusset plates at joints, which resulted in stress risers— bad news. Cracks happened, but that was the way it was. These cracks had to be welded up as part of operations. A better way was needed;

4. The Offshore Company (TOC) was better in its designs of leg joints, but the bracing to chord joints still had difficulties with stress concentrations that led to cracks;

5. Cast steel was used in TOC jackups for the pin hole rack castings, which were welded to the tubular chords and received the elevating pins in square rack holes. This meant that there was some precedent for the use of cast steel components in offshore service and in the structure of jackup legs;

6. One of the great things about cast steel joints was that the weld with the leg brace member (which in itself could be a stress raiser) could be made away from the high-stress locations at the leg chord intersection. Material could be made thicker to suit local loadings. Radii at joints could be chosen to mitigate stress concentration.

7. In addition to structural efficiency, we felt there was a simplification in fabrication as tubular structural members (pipes) could all have simple end cuts with no diagonal or saddle cuts;

8. Despite all these reasons, people in 1973's offshore world were nervous about using steel castings because they thought they would likely crack or be very expensive. My counterargument was that steel castings were used for high pressure steam lines inside nuclear submarines; there could hardly be a more demanding application! Making standard castings by the hundreds for ETA jackups would also produce savings from volumes production;

9. Steel castings in leg joints, both at chords and at the horizontal K brace point, were key in reducing leg weights and keeping average stress levels higher so that more load could be carried with less steel;

10. A byproduct of the structural efficiency achieved with cast steel joints was the

reduced need to take leg sections off for ocean moves. In the days before jackups could be conveniently put onboard a heavy lift vessel with lower ocean transit motions, this was huge;

11. Floating stability was not too difficult to achieve for the ETA jackup designs, partially because of the lightness of the legs from their cast steel joints.

In retrospect, our timing for the introduction of cast steel joints turned out to be good.

Accident rates with jackups in the early 1970s had reached high levels around tenpercent annually, as Table C indicated earlier. The industry was aware of this and tried to address it. In December of 1976, the Southwest Research Institute (SWRI) in San Antonio completed a study for Marathon Marine Engineering. To investigate fatigue and stress concentration levels in the leg chords in LeTourneau jackup legs, which were particularly susceptible to these problems. Their fracture mechanics analyses used freshly developed finite element methods to work the problem.

Around this time, the first ETA-designed jackups with cast steel joints were delivered in France and Japan. Two to four years later leading Japanese fabricators would market their own cast steel nodes for jackups (discussed more in Part V).

Cast steel joints by no means were a traditional building block for steel fabricators, whether for offshore or onshore oil and gas business— or anywhere in civil engineering. In hindsight, more time should have been taken in discussing the adaptation of fabrication methods to use steel castings. We saw a good engineering case for cast steel joints but did not develop a thorough business-oriented case. We should have had a cost-benefit analysis with comprehensive explanations such as:

(a) A comparison of weights and delivered costs for a conventional leg versus a version with cast steel joints for the same operating conditions:

(i) Fatigue loadings,

(ii) Survival loadings on location,

(iii) Loadings during ocean tows, when the legs may be oscillating from one side to the other, and

(iv) Impact loads going on location, when the legs touch bottom with the hull moving around in the waves;

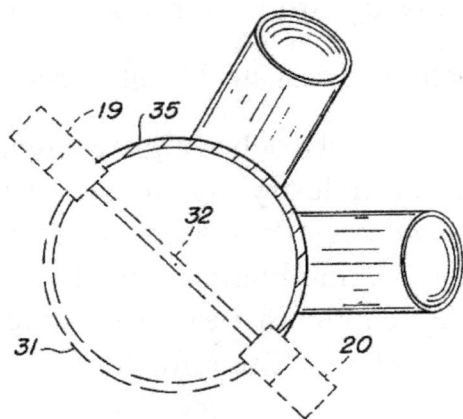

FIG. 8

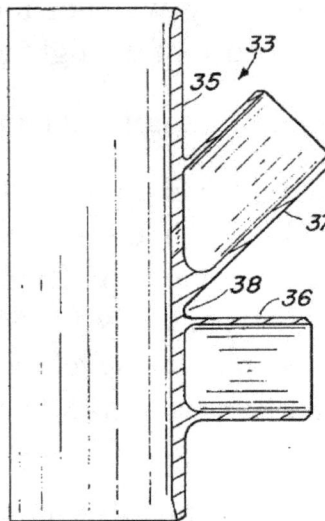

FIG. 9

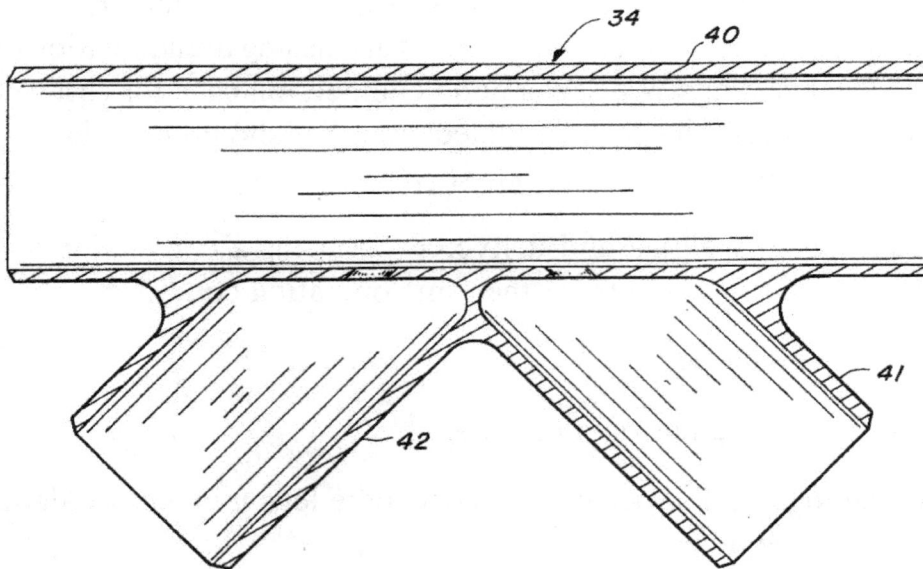

<u>**Diagram 17**</u>:
Patent detail of ETA's cast steel joints for leg chord and K brace connections

A comparison of fabrication plans and jigs, using castings from well-qualified and established foundries to think through what was really doable with jigging and tolerances with both leg types and the economies associated with these arrangements.

Unknown to ETA, there had been offshore precedent for the joint idea in the construction of the *Chris Chenery* semisubmersible designed by The Offshore Company, which circa 1972 specified forgings at the intersections of major structural members.

The offshore world has employed cast steel joints after ETA's time in a few production platforms and application in the GVA designs of semisubmersibles. Castings in these semisubmersible MODU structures can only be at a handful of tubular joints, unlike the hundreds of joints needed for a jackup with three long truss legs, which can then be repetitively produced with much bigger economies.

As the first ETA designed jackups with cast steel joints entered service, leading Japanese shipyards advertised their capabilities to build cast steel nodes via technical brochures, NKK in 1976, Hitachi in 1978 and Kawasaki in 1979. NKK ran a strain gage test to show the advantages of using their joint connection technology and published the results in a technical article in *Oil & Gas Digest* in September of 1980 – more about all this in Part V. In recent years, more technical papers have been written about the feasibility, features, and performance of cast steel joints in civil engineering structures, quite apart from offshore applications.

ETA was ahead of its time in calling for cast steel joints in these jackups.

ETA's 1975 Design Drawings For The Cast Steel Joints On The ETA Europe Class Jackups

Diagrams 18 and 19 are scans of the original ETA design drawings from 1975 for both the K bracing connection and the chord connections employed in the most demanding of the ETA jackup designs, the ETA Europe Class.

They show the practical details corresponding to the principles shown in the patent and discussed above. To my knowledge, this is the first time that design details of these jackup joint castings have been published. They are shown here in landscape mode instead of the portrait style and occupy the full page so that the drawings can be as readable as possible to see the increased wall thicknesses and radii used at the intersections of tubular members, shown in:

These diagrams also show dimensions and tolerance. It becomes possible to see how the cast steel joints look more like piping joints rather than a new and special structural element!

Reading these drawings in a printed version of this document may require good eyesight! If one views an electronic version of these diagrams on a computer screen, blown up at, say, 200%, then all the dimensions and notes are more readable. Alternatively, printing the pages with drawings on 11x17 in. paper would have the same effect.

The actual castings are relatively simple. Materials and tolerances were not difficult to achieve. The castings are generally similar to what is used in the power industry for high pressure steam piping. ETA was just adapting the idea for a simple structural only application with no internal pressure or high temperature requirements.

One might therefore conclude that the basic design concept is not original, and often used in far more critical service!

Part V gets into the realities of construction history with the different designs from ETA with pictures of the steel joint castings installed in the jackups built by Hitachi in Japan and by CFEM in France.

Diagram 18 (Next Page):
1975 design drawing of K brace steel castings in the ETA Europe Class jackup leg
Source: ETA Drawing *"Leg K Bracing Casting Details—27 Bays w/ ETA Jacking Unit" Dwg. No. 00630-064, dated May 14, 1975, Revision 3*

Diagram 19 (Page 124):
1975 design drawing of leg chord steel castings in the ETA Europe Class jackup leg
Source: ETA Drawing *"Leg Chord Casting Details—27 Bays w/ ETA Jacking Unit" Dwg. No. 00630-065, dated May 14, 1975, Revision 3*

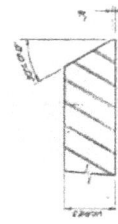

TYPICAL NONSPECIFIED TOLERANCES

Dia.	ACROSS PARTING LINE	BETWEEN POINTS FORMED BY CORE 1 WALL	BETWEEN POINTS IN ONE PART OF MOLD
1-7 IN.	.050	.050	.030
8-12 IN.	.060	.040	.050
13-16 IN.	.070	.070	.050
17-30 IN.	.080	.080	.080

NOTE:
MAT'1 TYPE MEMBER 'Y' DESIGNATE ASTM A148 (FORMED) STEEL
MAT'2 TYPE MEMBER 'X' DESIGNATE ASTM A148 (43,145) STEEL

LEG K-BRACE CASTING SCHEDULE

CASTING NUMBER	MATERIAL TYPE (SEE SPEC'S)	CORE VARIATION			QUANTITY FOR ONE LEG
		NUMBER	X DIM	Y DIM	LOCATION	

CASTING NUMBER	MATERIAL TYPE	NUMBER	X DIM	Y DIM	LOCATION	QUANTITY FOR ONE LEG
75-1-1	1	1	10.000"	10.500"	27, 26, 25, 24, 23, 22, 21, 20, 19, 3, 2, 1, 48	36
75-1-2	1	2	11.000"	10.500"	16, 15, 14, 13, 12	5
75-2-3	2	3	11.250"	11.250"	16, 15, 14, 13, 12	15
75-2-4	2	4	11.500"	11.500"	11, 10, 9, 8, 7, 6, 5	21
75-2-5	2	5	11.000"	11.000"	4	3
75-2-6	2	6	11.250"	11.000"	17	3

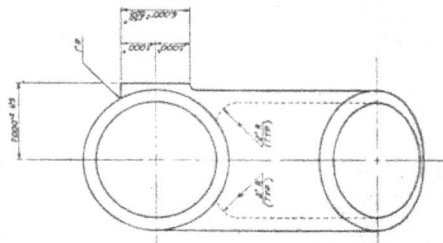

DETAIL 'J'
SCALE FULL

SECTION-B-B

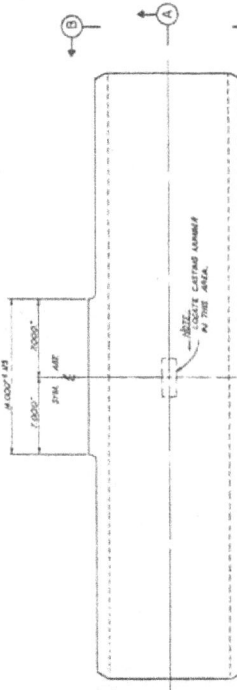

PLAN
LEG K-BRACE CASTING

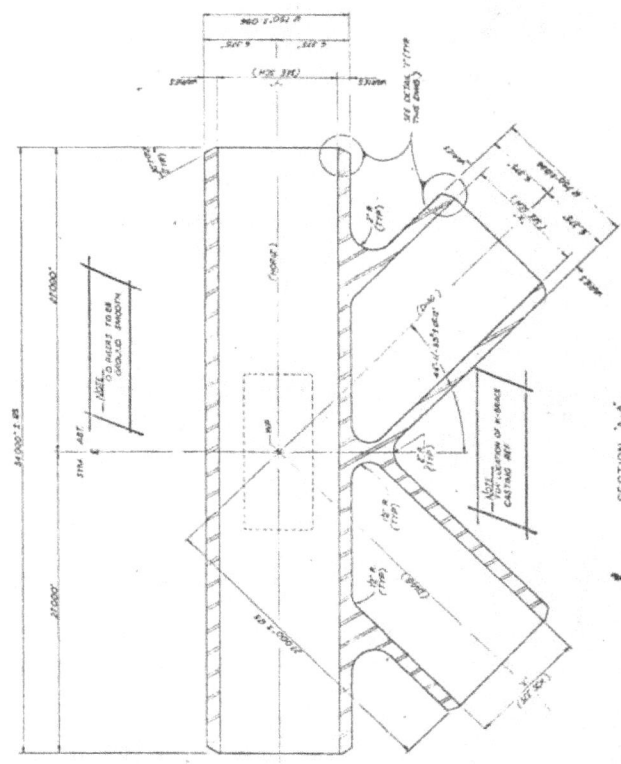

SECTION A-A

Engineering Technology Analysts, Inc.

THE ETA EUROPE GLASS.
LEG K-BRACE CASTING DETAILS
22 BAYS WFTA JACKING UNIT

DRAWING NUMBER 09630-064   REV 3.

REVISIONS

TYPICAL NONSPECIFIED TOLERANCES

DIM.	ACROSS PARTING LINE	BETWEEN POINTS FORMED BY CORE & WALL	BETWEEN POINTS IN ONE PART OF MOLD
1-7 IN.	.050	.050	.030
8-12 IN.	.060	.060	.040
13-16 IN.	.070	.070	.050
17-20 IN.	.080	.080	.060

LEG CHORD CASTING SCHEDULE

CASTING NUMBER	"Z" DIM	MATERIAL TYPE SEE SPEC.	CORE VARIATION				QUANTITY ONE LEG
			NUMBER	"X" DIM	"Y" DIM	LOCATION	
TG-1-1	13.375	1	1	10.500	16.500	21,16,25,24,23, 22,20,20, 19,21,0	29
TG-1-2	13.375		2	11.250	11.250	46,53,14,9,12	13
TG-1-3	13.375		3	11.500	11.500	11,10,9,8	12
TG-1-4	13.375		4	11.750	11.750	7,6,5	9
TG-1-5	13.375		5	11.000	11.000	17	3
TG-1-6	13.375		6	10.000	10.500	3	3
TG-1-7	13.375		7	11.750	11.000	4	3

PLAN — LEG CHORD CASTING

SECTION 'B-B'

SECTION 'A-A'

DETAIL '1' SCALE 3"=1'

DETAIL '2' SCALE 3"=1'

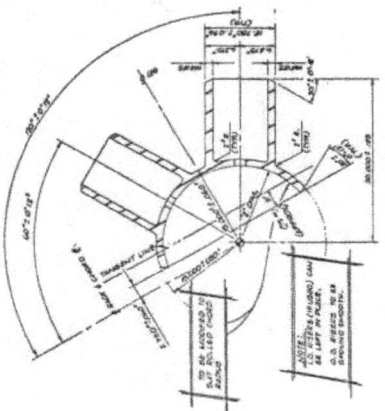

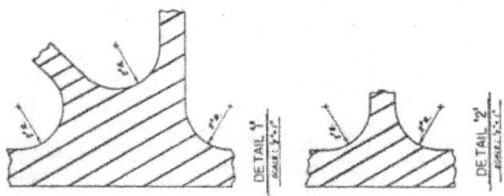

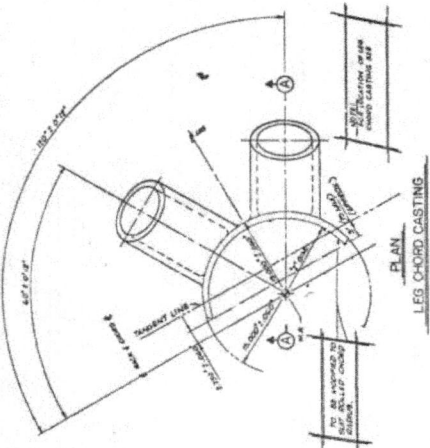

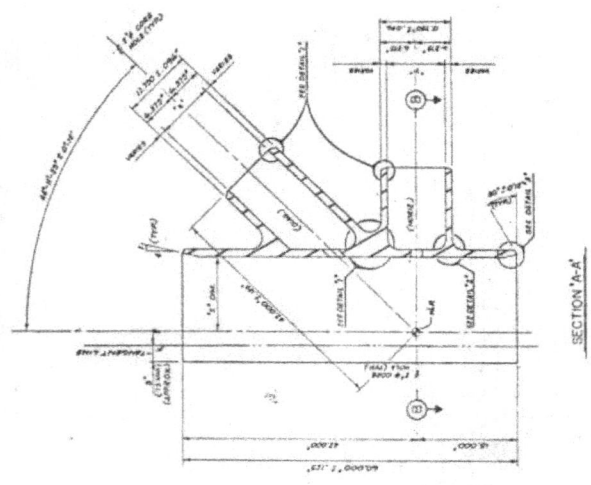

THE ETA EUROPE CLASS.
LEG CHORD CASTING DETAILS
27 BAYS W.E.T.A. JACKING UNIT

DRAWING 09630-065

Engineering Technology Analysts, Inc.

REVISIONS

ETA's 1975 Design Drawings For The Legs On The ETA Europe Class Jackups

The drawings of the cast steel joints are followed by two more original ETA design drawings from 1975 of the ETA Europe Class for the leg design. Diagrams 20 and 21 are two more 1975 design drawings for the legs which show details of the leg design, overall leg construction, and use of the cast steel joints.

Forty year ago, these two drawings were highly proprietary information! And they were displayed usually in 22 x 34 inch prints instead of book size illustrations. Today they are interesting and hitherto unavailable history:

Once again, dimensions and notes are visible, along with edge preparations for welding.

Diagram 20 (Page 126):
Leg design details, arrangement used in the ETA Europe Class jackups
Source: <u>ETA Drawing</u> *"Leg Assembly Sheet 1—Bays w/ ETA Jacking Unit" Dwg. No. 00630-061, dated February 19, 1975, Revision 3*

Diagram 21 (Page 127):
More 1975 leg design details in the ETA Europe Class jackups
Source: <u>ETA Drawing</u> *"Leg Miscellaneous Details—27 Bays w/ ETA Jacking Unit" Dwg. No. 00630-063, dated May 14, 1975, Revision 3*

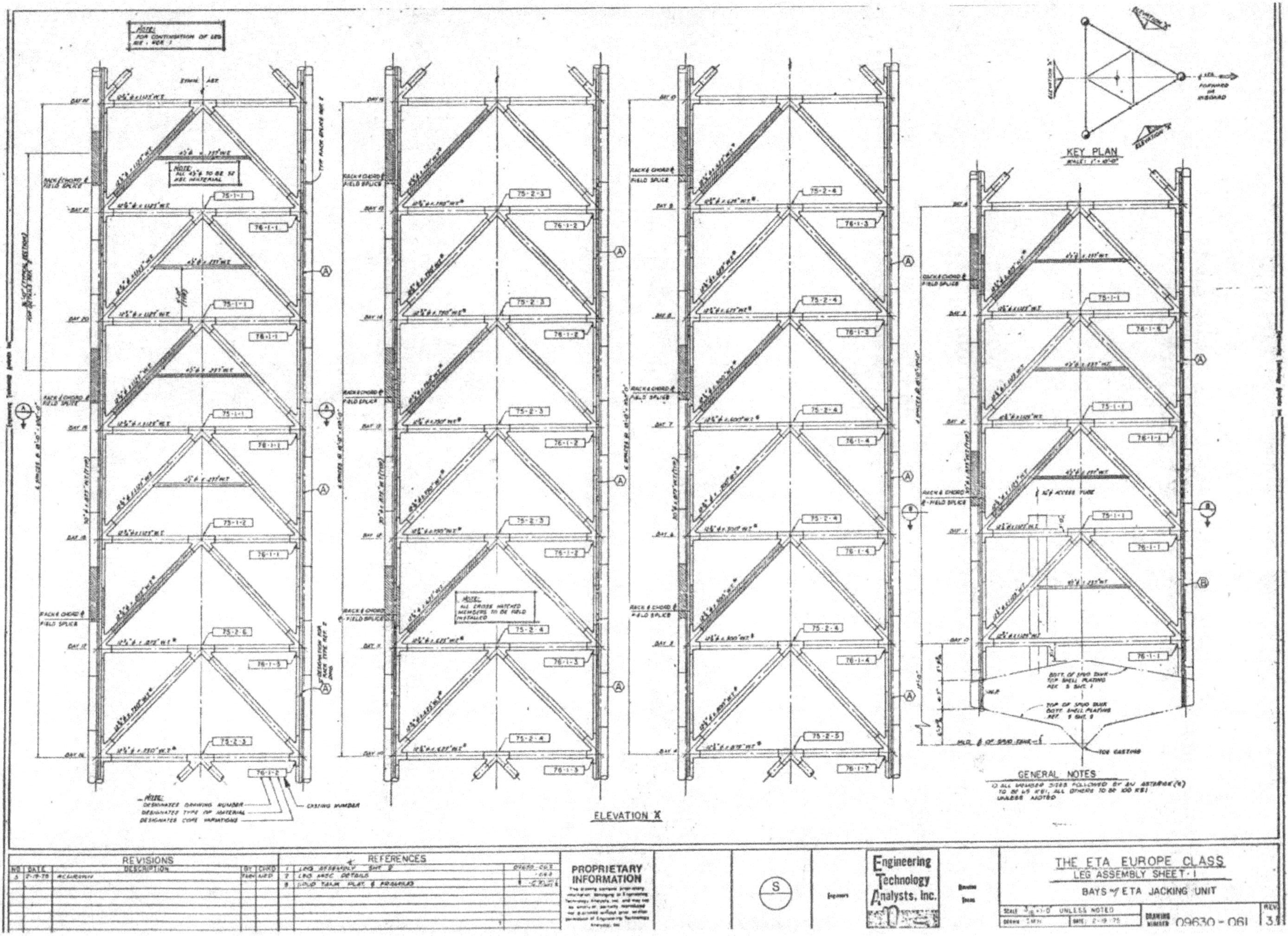

ELEVATION X

KEY PLAN

GENERAL NOTES

THE ETA EUROPE CLASS
LEG ASSEMBLY SHEET 1
BAYS of ETA JACKING UNIT

DRAWING NUMBER 09630 - 061

Engineering Technology Analysts, Inc.

PROPRIETARY INFORMATION

REFERENCES

REVISIONS

ETA's Rack And Pinon Elevating System

ETA's designs had started out to use the rack and pinion elevating system from National Supply, manufactured in the United States. It was used for the ETA Robray 300 Class jackups, and was initially proposed for the ETA Europe Class jackups to be built at CFEM in France.

CFEM preferred for the jacks to be manufactured in France so that they could be included in the overall COFACE (Compagnie Française d'Assurance pour le Commerce Extérieur) project financing plan to increase exports from France. COFACE has been the export credit agency for France since its foundation in 1946. It is likely CFEM also had their eye on capturing the jacking system business in France for themselves or at least another French company.

This led ETA to create its own rack and pinion elevating system by making an agreement with Lufkin Industries, a well-established supplier of oilfield equipment, based in Lufkin, Texas. ETA developed its own elevating system design, derived from a system previously manufactured by Lufkin Industries. A few years earlier, Lufkin had used this system in a different configuration for three six legged Transocean jackups (no relation to the modern Transocean) that used tubular rather than truss legs. This Transocean was a predecessor of Transworld Drilling that was later acquired by Noble Corp. In 1965, the *Transocean 1* jackup was delivered from Howaldtswerke Deutsche Werft (HDW) in Kiel, Germany. It was followed by two more jackups delivered in 1973 and 1976: *Transocean 3* and *Transocean 4*. Thus there was some practical history behind this rack and pinion elevating system proposed for the two ETA Europe Class jackups.

I therefore negotiated a licensing agreement for Brissoneau et Lotz Marine (BLM) in France to build ETA's system, and that carried the day for CFEM and the COFACE financing of the two jackups. Accordingly, ETA adapted the design of the ETA Europe Class jackup to employ the new ETA rack and pinion system for the two jackups to be built for Dyvi Drilling A/S at CFEM's yard in Dunkerque.

One of the features of this rack and pinion system was that each pair of opposed pinions was set in its own cell in the jackhouse structure with layers of steel and rubber for shock absorption purposes. This contrasted with the National Supply jacks that were supplied as a single two or three high elevating unit, all in a single structure. These National Supply jacks still had shock absorption layers but for the complete unit. We felt the ETA design offered an edge in shock absorption.

With their unique configuration, ETA jacks have performed very well in the North Sea. Their success shows how the typical jackup impact effects were much decreased while going on location, experiencing storm waves, and being jacked up. The theory behind the design had predicted that, but it has been satisfying to see. The ETA legs were safer than expected.

After ETA went out of business in 1977, BLM built more jacks on their own, including sets for the ETA Robray 300 Class jackups for India's ONGC (Oil & Natural Gas Corporation) which Robin Shipyard delivered in 1982, and for *Pool Arabia 145* which FELS delivered in 1982.

Years later, Hydralift acquired BLM. In 2014, the Houston-based manufacturer National Oilwell Varco (NOV) acquired Hydralift, so today NOV offers both the original ETA and National Supply designs of rack and pinion elevating systems.

After their experience with ETA's jacks in the *Dyvi Beta* and *Dyvi Gamma*, CFEM developed their own elevating system for subsequent use in their own jackup designs.

Rack Chocks And ETA Jackups

To avoid having the rack and pinion system take all the loads on location under their holding load criterion, it is desirable to have a leg-locking system or *rack chocks*. The principle was that the allowable loading taken while on location and jacked up could be increased by installing a locking system.

Loads due to rig overturning were then transferred directly from the jackup hull to the legs. Thus the legs could withstand higher storm-induced design loads. The effect was that we avoided the need for additional pinions when holding loads would have demanded the significant expense of adding pinions to cater for holding loads only.

The ETA Europe Class design was not originally designed with rack chocks but had a leg-wedging system (described in the operating book for *Dyvi Gamma* dated April 26, 1977).

Rack chocks were designed in 1976-1977 and considered for the ETA Europe Class design but ultimately not used. Dyvi Drilling A/S was attempting to qualify one of their ETA Europe Class jackups for a location in the North Sea. That location was several degrees north of Ekofisk in the UK sector in Scottish waters, where the metocean conditions were beyond allowables for *Dyvi Beta* or *Dyvi Gamma*. Accordingly Ramesh Maini, who had newly joined Dyvi Drilling, and Mike Guidry, vice president of operations with Dyvi, designed a leg-locking system that allowed the ETA Europe Class design to qualify for the proposed location.

The London office of the Houston-based Lawrence Allison prepared the engineering. Ramesh Maini had been ETA's representative at the building yard (CFEM) and had an excellent grasp of the loadings and design particulars from his earlier work on both the ETA Robray 300 Class and the ETA Europe Class jackups.

Ultimately the two Dyvi jackups were assigned to other locations, and there was no need to proceed with the rack chock installation.

Dyvi Drilling was in a position to file a patent application on the rack chock design but did not do so. Furthermore, in 1976-1977, ETA management seemed to not understand the significance of this advance; ETA made no attempt to collaborate in a patent to protect this intellectual property. The upshot was that neither ETA nor Dyvi Drilling took the necessary action to secure a proprietary position on the rack chocks design. This was unfortunate because, just a few years later, other jackup designers developed and patented locking systems, most notably Friede & Goldman with their L780 series of designs, later followed by separate patents by Keppel, MSC, Zentech, and BLM. In its 1983 technical brochure, Hitachi marketed an "Independent Tooth Type Leg Clamping System." ETA had missed out on the opportunity to secure rights to what became an industry standard feature.

In 1991-1992, the conflict around patent rights of rack chocks was tested when another ETA-designed jackup was modified to operate in the North Sea as an accommodation unit. The *Norbe I*, an ETA Robray 300 Class jackup owned by Odebrecht of Brazil and delivered in Singapore by Robin Shipyard in 1979, was acquired by Capital Maritime and modified by Bethlehem Steel in Beaumont, Texas with final completion at Breivik in Norway.

For engineering the conversion, its owner, Capital Maritime, had retained the services of Zentech of Houston, where Ramesh Maini was now president. Kvaerner from Norway was also involved in the design of the accommodation facilities and project and construction management. The classification of the unit changed from ABS to DNV. The legs underwent extensive modifications to enable the jackup to operate safely as an accommodation unit in the North Sea. In the fifteen years since the first

ETA jackup entered the North Sea, significant learnings and rigor had developed North Sea standards for jackups design (more on this topic in Part V).

In order to stretch the capability of the Robray 300 Class jackup to satisfy North Sea criteria, Zentech designed a rack chock system along the lines of the ETA Europe Class precedent. However, Capital Maritime would not accept the prior art design from the *Dyvi Beta* and *Dyvi Gamma* and chose to pay a license fee to Friede &Goldman to avoid conflict over patent rights on Zentech's design of rack chock system.

After conversion, that accommodation unit was known as *Rigmar 301* and as *Port Rigmar*. Today it is called the *COSL Rigmar*. Still in existence, it is believed to be laid up in Dutch waters.

The early pioneering of rack chocks for ETA jackups was not widely known in the industry. Today, the common impression is that Friede & Goldman first designed rack chocks, although they secured a patent position, some would claim that they were not the first! Forty years later, the debate may not make a lot of commercial and practical difference, but the true inventors need recognition!

ETA's Advances On Assessing The Floating Stability Of Jackups

It had become obvious to many that ETA was breaking new ground in jackup design. However, there was a more subtle development emerging as classification societies and underwriters vetted successive ETA designs for intact and damaged floating stability and readily accepted them.

Each vetting organization was very stringent in their approval of a jackup design for floating stability in both the intact and damaged conditions; no one wanted to risk the jackup ever capsizing while afloat because there had been such accidents. And these jackups with unusually long legs were designed by some new outfit!

ETA's ability to gain such approvals was no accident. It took a lot of trial and error and many late nights. We had discovered that applying common rules of thumb and industry practices to jackups led to potentially disastrous results for the kind of triangular-hulled jackups that ETA designed with long legs. The ETA Robray 300 Class jackups had legs 425-ft.-long, and the ETA Europe Class set records at legs that were 508-ft.-long.

Computer methods for floating stability calculations for jackups were still rarely used, so floating stability calculations were laboriously done by hand. The number of necessary calculations grew and grew. Better methods were needed.

The structural advances discussed earlier drew on to patented design features and quite specific load-carrying results based on structural models and experience in loadings and materials behavior. However, the naval architectural principles were built on generations of marine-operating experiences with ships. Lessons were hard-won from disasters. For example, there were advances in compartmentation and damaged stability after the loss of the *Titanic* in 1912.

The upshot from all the stability investigations in 1972-1973 was that two of ETA's experts got to thinking: how could we show the offshore community that there were better methods and that they should accept our new methods? We had a remarkable balance of talents and backgrounds—Ralph McTaggart, ETA's chief naval architect, who had several years of experience in stability assessment of jackups, and Dr. Richard "Dick" Gunderson, who had outstanding structural mechanics and analytical skills. Together they had tackled many of the stability analyses of jackups at ETA. They presented a paper at the Annual Meeting of the Society of Naval Architects and Marine Engineers (SNAME) in New York in November of 1973. It was a thorough piece which introduced more realistic methods for gauging floating stability.

Diagrams 22 and 23 that follow summarize their findings with extracts from that paper, which became the standard framework for assessments of jackup floating stability.

To understand some of the principles involved in this arcane world of floating stability and their application to jackups, these notes may help:

a. When a ship heels, the center of buoyancy of the ship moves laterally. It might also move up or down with respect to the water line. The point at which a vertical line through the heeled center of buoyancy crosses the line through the original, vertical center of buoyancy is the *metacenter*. The vertical distance between G and M is referred to as the metacentric height (GM). The relative positions of vertical center of gravity G and the initial metacenter M are extremely important because of their effect on the ship's stability.

b. The ship is in stable equilibrium if G is below M, in neutral equilibrium if the vertical center of gravity (VCG) and M are coincident, and in unstable equilibrium if VCG is above M. If the metacentric height of a ship is small, the righting arms that develop will be small. Such a ship is *tender* and will roll slowly, paradoxically being more comfortable for its passengers. However, if the metacentric height (GM) of a ship is large, the righting arms that develop at small angles of heel will be large. Such a ship is *stiff* and has more resistance to rolling. The metacenter remains directly above the center of buoyancy by definition. The small Fig. 3 at the top right of Diagram 22 shows how that works.

c. In a jackup, the center of gravity is far above the center of buoyancy, often several times the height of the center of buoyancy from the keel. In the pro forma analyses in Diagrams 22 or 23, KG > KB and KG was 50 ft. and KB was 6.5 ft.. And yet jackups can still float stably if the conditions are right. Fig. 8 and Fig. 9 show how the value of KG makes a huge difference in safe floating stability, whether intact or damaged and why this factor must be very carefully calculated. Further, it is a factor that the barge master responsible for loading and moving the jackup will be easily able to compute, i.e. determining that KG value is doable in the field rather than at the office.

d. A jackup is much wider than a ship and can have a ratio of length to beam (L:B) of about 1.0 while a ship can be several times that. (VLCCs commonly have L:B of around 5.5.) For damaged stability calculations, heel is the dominant effect on a ship since the length is much more than the beam. However, for a jackup that is triangular and about the same length as beam, the trim effect (heel in the bow-to-stern direction) can be critically important for a damaged stern compartment, e.g. the location indicated in Fig. 6 in Diagram 22.

e. The industry convention for damage stability is that the wind heeling effect exists with a 50-knot wind (assuming still water and calculation for heel with that 50-knot steady wind). For intact stability, the wind is taken to be more, which raises the question of what wind speed should be used. This has an important effect on the allowable leg length: the more wind, the less leg length can be safely extend above the hull, as illustrated in Fig 8 and Fig.9 at the left of Diagram 23. Lowering the legs for a tow reduces the KG and improves stability.

f. The industry requires factors of safety meeting 1.05 for damaged stability and 1.40 for intact stability. These measurements are computed from the ratio of areas under the righting moment curve (see several of the Figures in Diagram 23) divided by the area under the overturning moment curve (i.e. the wind heeling effect).

g. The shift to using KG as the key metric had a profound practical effect. Once the basic behavior of the jackup could be characterized as in the Figures in Diagram

23, calculation of KG could be done simply and quickly and checked against the graphs of the safe KG needed for damaged and intact stability. No longer did it take home-office naval architects!

h. The debates still continued on policy matters as to whether a 50-knot wind and a 1.05 factor of safety were indeed safe enough for a damage condition and whether a 1.40 factor of safety in combination with a 100-knot wind was too safe for the intact condition. But now the complexities and time for calculations were easier.

This quick outline does not attempt to get into all the theories of floating stability but simply to highlight the serious differences that the ETA investigation found to be important.

Floating stability was a particularly critical issue in the early 1970s. Five to ten years later, the idea of *dry tows* became common. In dry tows, jackups would be loaded on a heavy lift ship, which then transports the jackup to its destination, perhaps halfway around the world. Previous to that, jackups were moved by towing them like any other type of floating barge or ship.

Dry tows were significantly faster at 12-14 knots compared to the wet tow's 3-4 knots. The risks of loss in long ocean tows were seriously reduced because dry tows were shorter time exposures and could often avoid storms. In addition, dry tows reduced the downtime for mobilizing from one location to another.

Thus when ETA disappeared in 1977, it left behind both visible advances in jackup designs and safety—plus an advances in how to make sure these jackups floated safely!

On Damaged Stability of Drilling Vessels

Ralph G. McTaggart,[1] Member, and Richard H. Gunderson,[2] Visitor

Intact and damaged stability of vessels with KG values greater than KB is discussed herein. A self-elevating mobile drilling unit, being a good example of a vessel with a high center of gravity, is used as an illustration. An example is given with calculations based on GM equal to BM showing the vessel to be stable when, in fact, it is extremely unstable. Also, KG is proposed as the indicator of a vessel's stability rather than the GM. The authors show that methods used in damaged stability calculations of conventional vessels do not apply to vessels with large KG values such as found in drilling vessels, particularly when the damaged condition produces large trim and heel angles. To calculate properly the stillwater conditions, the shift in longitudinal center of buoyancy due to trim must be included for vessels with KG values greater than KB. Completing damaged stability calculations requires several computational steps utilizing hydrostatic data based on the damaged ship in both the trimmed and heeled position.

[1] Chief Naval Architect, Engineering Technology Analysts, Inc., Houston, Texas.
[2] Senior Engineer, Engineering Technology Analysts, Inc., Houston, Texas.

Presented at the Annual Meeting, New York, N. Y., November 15-17, 1973, of The Society of Naval Architects and Marine Engineers.

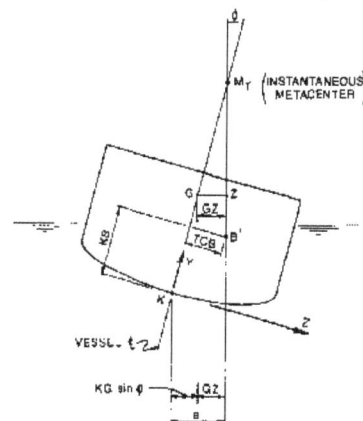

Fig. 3   Transverse righting arm and center of buoyancy

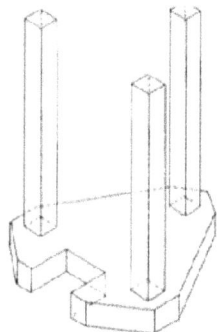

Fig. 1   Typical self-elevating drilling unit

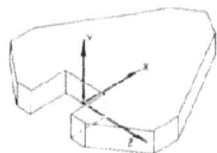

Fig. 2   Coordinate system for the vessel

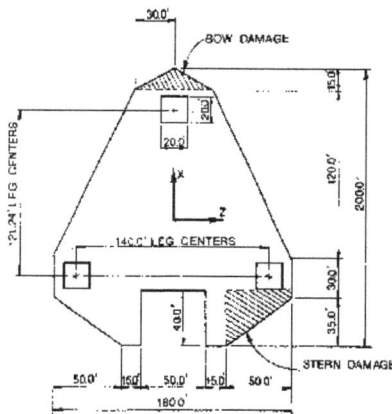

Fig. 6   Plan view of self-elevating drilling unit showing damaged portions

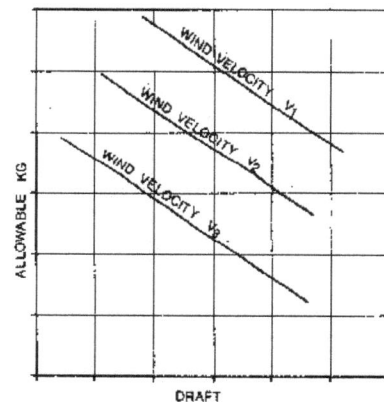

Fig. 14   Allowable KG curve

Diagram 22:

Summary, basis for ETA's investigation of floating stability of jackups

Source: SNAME paper. November 1973

Fig. 8 Stability curves-- no damage

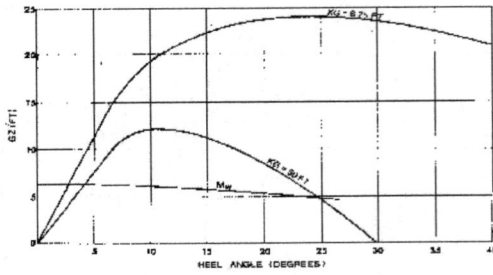

Fig. 9 Stability curves— bow damage

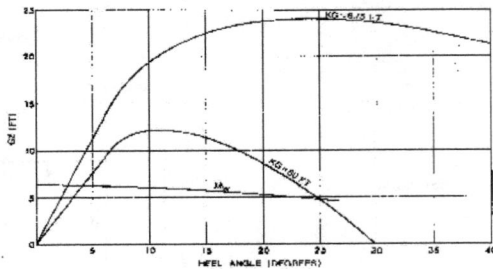

On Damaged Stability of Drilling Vessels

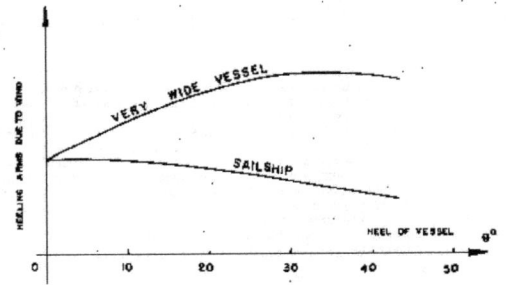

Fig. 10 Stability curves— stern damage

Fig. 19 Wind heeling arms diagram

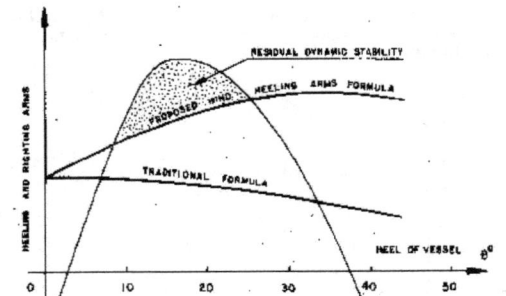

Fig. 20 Stability diagram in damaged condition

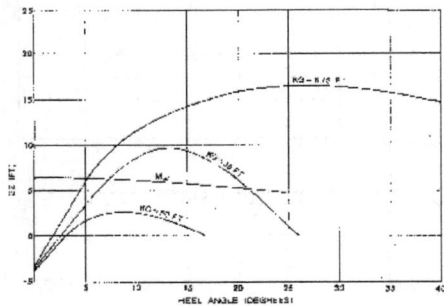

Diagram 23:
Summary of results from ETA's investigation into floating stability of jackups,
presented to the industry in November 1973
Source: SNAME paper, November 1973

PART IV

BRAINSTORMS, CONCEPTS, AND PATENTS: PAPER ONLY!

THE JACKUP WITH ONE LEG THAT INSPIRED OTHERS

Brainstorming new ideas was a fun part of what we did in ETA. Part IV is about the new ideas that we felt were worthwhile at the time, even brilliant, but in the real world never made it past the paper stage.

When we brainstormed, we wrote up what we felt were the better concepts and applied for patent coverage in the United States. Ultimately four patents were granted but never used; they are described here to show the problems brainstormed and the solutions created.

These concepts were intended for use in new jackup designs in the hope that they would have fundamental value and prompt interest among potential clients. We always had to strike the balance between coming up with something nutty or something that might have some future, even something future generations would say was a "game changer!"

Reactions from potential customers helped directions for specific design advances. Reactions taught us a lot as a kind of test marketing. It was stimulating to dream up these ideas. We really did not see it as engineers playing at their profession or being mad inventors!

It took some time for us to learn to dream up concepts that were close enough to existing methods and sizes so that clients would be enticed to take the risk and try them with a good probability of payoff. One of our early concepts was far too far out.

The ETA Deepwater Jackup

ETA's first new jackup design concept was the slant leg "ETA Deepwater Jackup." It publically offered a 400-ft. water depth capability, although internal design studies showed the leg and hull sizes could be used in up to 500-ft. water depths. It was first introduced in the January 10, 1972 issue of *Oil & Gas Journal* and was pictured in the artist's impression on the cover of *ETA Innovation* (see Picture 1), published in the first quarter of 1972.

An illustration in the April 1974 *Northern Offshore* article showed the ETA Deepwater Jackup while afloat, reproduced here in Picture 11.

Picture 11:
The ETA Deepwater Jackup afloat
Source: "Jackups – A Future in the North Sea." Northern Offshore, *4th quarter*

The consumables capacity for the ETA Deepwater Jackup was relatively large, roughly twice what was typical for a 300-ft. jackup of that era. It was similar to what would be done with the ETA Europe Class design and the ETA Mobile Monopod, which were developed two years later. Table E is extracted from a 1974 article in *Offshore* magazine and cites the consumables figures.

Consumables capacities (Typical for Drilling Operations)	Europe class	Deepwater Jackup	ETA Monopod
Drilling water, barrels	7,380	6,000	6,100
Potable water, barrels	1,290	900	850
Diesel fuel, barrels	4,870	4,500	4,500
Dirty oil, barrels	260	100	100
Liquid mud, barrels	2,800	2,800	2,800
Bulk mud (4-12 ft. diameter tanks), cu ft.	4,520	4,520	4,520
Bulk cement (4-12 ft. diameter tanks), cu. ft.	4,520	4,520	4,520
Deck pipe rack capacity, lbs.	1,000,000	900,000	900,000
Sack storage, sacks	6,000	6,000	6,000
Additional consumables capacity unassigned, lb.	1,460,000	403,000	388,000
Total consumables capacities, lbs.	9,500,000	8,500,000	8,500,000

Table E:
Big variables capacities (for 1974) specified in the new ETA jackup designs
Source: "New Innovative Designs for the North Sea Oilpatch" <u>Offshore</u>, *April 1974, 4 pages*

Back when ETA made its November 1972 proposal to Marathon LeTourneau for developing this design, quite a number of operating conditions were assessed in these early days (see Diagram 24) beyond what made it into print.

Extracting data from the old, faded, and rather difficult-to-read proposal allows the assembly of a set of design criteria for the ETA Deepwater Jackup's capabilities in water depths up to 500 feet. Diagram 24 shows a total of 10 sets of design criteria. For the purposes of the discussion here, only 4 of these are cited here in Table F for water depths of 250, 400, 450 and 500 ft. Two sets of criteria are shown for comparison with *Zapata Nordic*: water depths of 278 and 306 ft.

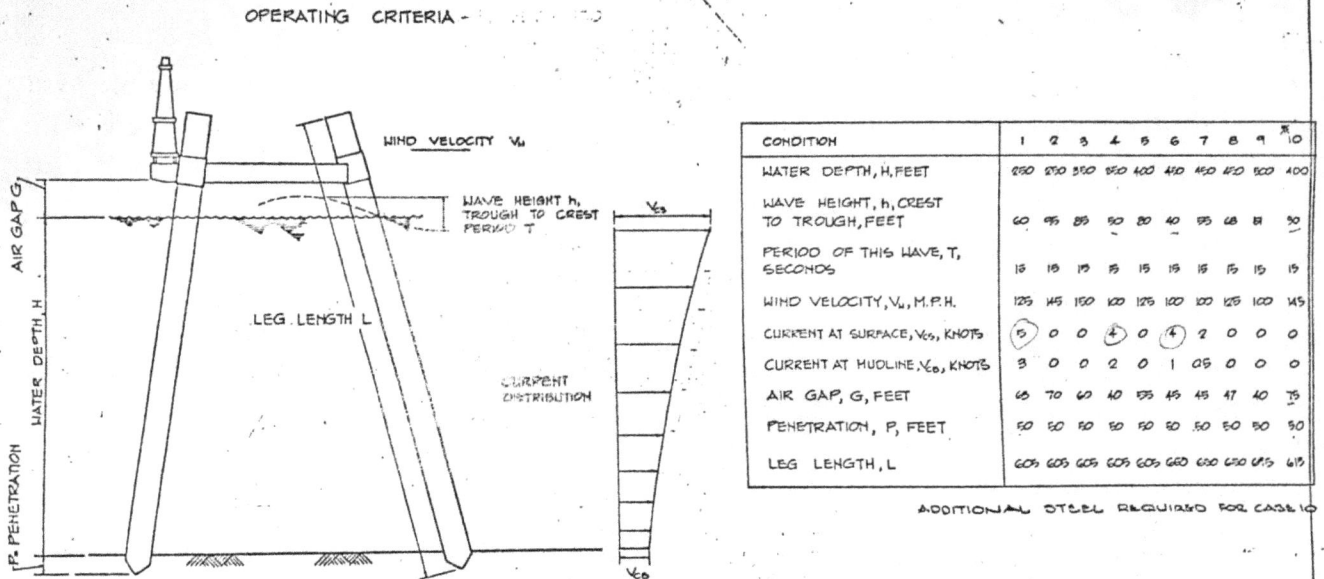

OPERATING CRITERIA -

CONDITION	1	2	3	4	5	6	7	8	9	10
WATER DEPTH, H, FEET	250	250	350	550	400	410	450	470	500	400
WAVE HEIGHT, h, CREST TO TROUGH, FEET	60	95	85	50	80	40	55	68	11	90
PERIOD OF THIS WAVE, T, SECONDS	15	18	15	15	15	15	15	15	15	15
WIND VELOCITY, V_w, M.P.H.	125	145	150	100	125	100	100	125	100	145
CURRENT AT SURFACE, V_cs, KNOTS	5	0	0	4	0	4	2	0	0	0
CURRENT AT MUDLINE, V_cb, KNOTS	3	0	0	2	0	1	0.5	0	0	0
AIR GAP, G, FEET	68	70	60	40	55	45	45	47	40	75
PENETRATION, P, FEET	50	50	50	50	50	50	50	50	50	50
LEG LENGTH, L	605	605	605	605	605	660	690	690	695	615

ADDITIONAL STEEL REQUIRED FOR CASE 10

Diagram 24:
Design criteria for ETA Deepwater Jackup
Source: ETA proposal *dated November 7, 1972*

The ocean tow criteria were typical for the ETA Deepwater Jackup; in the ocean tow calculations, a 20 degree roll off center was assumed to determine the amount of leg to be removed, i.e. for operation in 500 ft. of water, a modest 90 ft. had to be removed from 695-ft.-long legs.

The leg material employed steel of up to 85 ksi yield. The hull was straightforward and conventional leg construction was little different from other jackup construction of the day. Drilling equipment, systems, hook loads, and living quarters were all typical for exploratory drilling in the industry. The consumables capacities were somewhat higher than typical, as the center column in Table E indicates.

The hull dimensions of the ETA Deepwater Jackup were 283 x 253 x 34 feet. Leg centers fore and aft were 165 ft. and 190.5 ft. between the two aft legs, i.e. very large, even by today's standards.

To put all this into perspective, Table F compares the largest slant leg jackup ever built, contrasting the ETA Deepwater Jackup with the *Zapata Nordic* delivered from LeTourneau's building yard in Vicksburg, Mississippi in March of 1972 as their hull no. 54. *Zapata Nordic* was designed for up to 306 ft. water depth with legs 440 ft. long. Its hull measured 234 x 233 x 26 ft. Leg centers fore and aft were 149 ft. and 172 ft. between aft leg centers.

The largest slant leg jackup ever built was recognized as a brave industry advance, yet its main characteristics were significantly smaller than the ETA Deepwater Jackup. Materials used in its construction and the drilling equipment were generally similar to ETA's. Rig-moving criteria were generally similar, but wind and wave storm criteria were somewhat less

ETA had no corporate ego trip and no intent to design something radically bigger and better. Rather, it was a matter of picking a good configuration for 400 feet of water that might be a future market need. We discovered we had picked something that could perform much better than expected.

We had gone too far. The figures in Table F show how the light ship (weight of hull, legs, and all equipment and systems) was more than twice that of *Zapata Nordic*. Roughly speaking, that meant the investment would be more than double to build! It was a big move for Zapata to go for the *Zapata Nordic*, so how could anyone ever go for an ETA Deepwater Jackup, which was so much bigger?

Looking back, in hearing ETA's proposal in November 1972, Marathon LeTourneau must have been (understandably) skeptical and critical: ETA had a heck of nerve to come to the leader in slant leg jackup design with this design concept for such a

Water depth	Wave height	Wave period	Wind velocity	Air gap	Leg penetration	Leg length	Single amp. roll	Leg removal, ocean tow
feet	feet	seconds	mph	feet	feet	feet	degrees	feet

March 1972:   *Zapata Nordic* delivered (LeTourneau hull no. 54), biggest ever slant leg jackup

	Aft leg centers, feet:		172	Leg centers fore and aft, feet:			149	
306	50	15	80	60	20	440	20	unknown
278	60	15	115	60	20	440	20	unknown

Hull dimensions, overall L x B x D, feet: 234 x 173 x 26
Consumables capacity, kips: 3,500 (estimate)   ) Hence 'light ship'
Total maximum weight on location, kips: 17,950   ) of about 6,652 tonnes
Equivalent to displacement in tonnes of: 8,152   ) displacement

November 1972:   ETA Deepwater Jackup

Design concept announced January 1972, proposed for further development 11/72

	Aft leg centers, feet:		190	Leg centers fore and aft, feet:			165	
250	95	15	145	70	50	605	20	0
400	80	15	125	55	50	605	20	0
450	55	15	100	45	50	650	15	45
500	51	15	100	40	50	695	8	90

Hull dimensions, overall L x B x D, feet: 283 x 253 x 34
Consumables capacity, kips: 8,500   ) Hence 'light ship'
Total maximum weight on location, kips: 41,190   ) of about 14,836 tonne
Equivalent to displacement in tonnes of: 18,706   ) displacement

Drilling equipment, drilling systems and hook loads were similar in both designs
Grades of steel used in both designs were similar

Table F:
Comparison of design criteria for the largest slant leg jackup in 1972 (*Zapata Nordic*) and the ETA Deepwater Jackup
Source: ETA proposal dated November 7, 1972

humongous unit—just eight months after they had delivered the biggest slant leg jackup ever! Whatever they thought internally, outwardly they were very polite and professional!

While the ETA Deepwater Jackup may have demonstrated creativity and a willingness to challenge boundaries, no client took it seriously enough to ask for further engineering. We had blown serious discretionary time and money in the design. More fundamentally, the theory may have been fine, but we learned that it was simply too big and didn't match what the industry needed, so there was little chance anyone would want it.

A few months later in our design for Robray, we were sure to align our designs with real-world needs.

Although never built, the ETA Deepwater Jackup concept served to gain industry attention for us as an upstart jackup designer! It signaled that ETA's marketing program was starting to work.

ETA's early structural analyses of jackups got us thinking about the operating conditions the industry really faced. This led to brainstorming ideas for proposing improved new jackup designs. Two such brainstorms for improving the ability to go on location are outlined next.

Going On Location: Mitigating Vertical Impact And Rolling, Two Patents Never Used

1. <u>Going on location with high-heave impact or punch-through hazards</u>. Sea conditions might create heaving. When a relatively flat spud tank hit the seabed, heaving could cause high-impact loads.

   Little was known then about punch through, but many feared that soil might give way under wave and direct loading, tilting the jackup hazardously. The proposed solution was a special design of high-penetration spud can to reduce high impact loads and penetrate more deeply into soil strata to reduce the risk of punch through. Diagram 25 shows the concept for ETA's pointy spud tank.

   <u>Patent</u>: *"Spud Tank for Offshore Drilling Unit," US patent no. 3,823,563, inventor: Peter M. Lovie, filed September 5, 1972, granted July 16, 1974.*

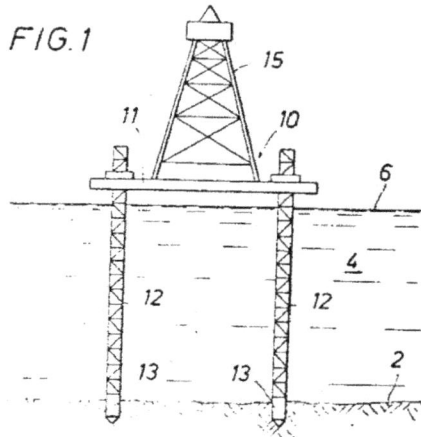

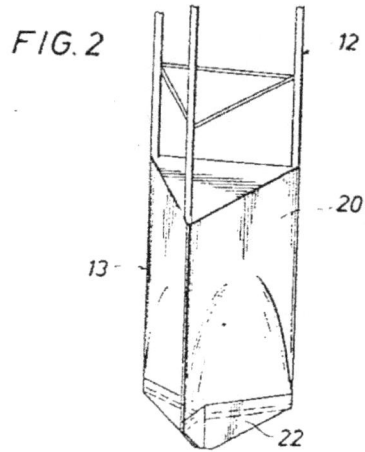

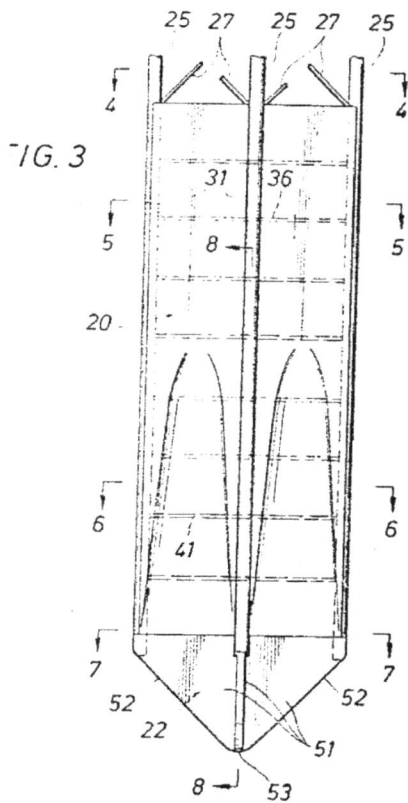

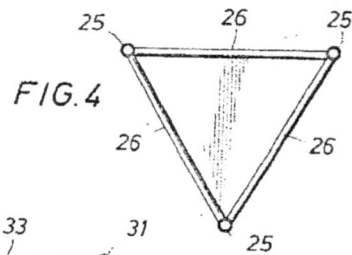

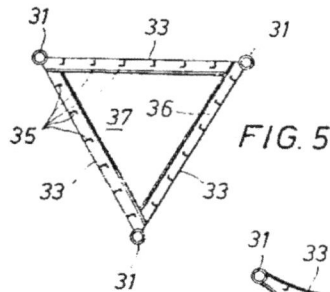

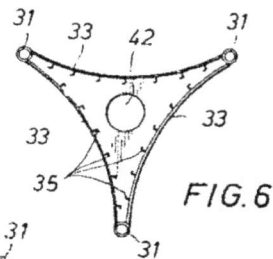

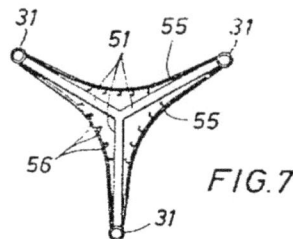

Diagram 25:
ETA's pointy spud tank concept

United States Patent [19]

Lovie et al.

[11] **3,916,633**

[45] **Nov. 4, 1975**

[54] **MEANS FOR ALTERING MOTION RESPONSE OF OFFSHORE DRILLING UNITS**

[75] Inventors: **Peter M. Lovie; Edwin L. Lowery,** both of Houston, Tex.

[73] Assignee: **Engineering Technology Analysts, Inc.,** Houston, Tex.

[22] Filed: **Aug. 24, 1973**

[21] Appl. No.: **391,070**

[52] **U.S. Cl.** **61/46.5;** 61/53; 37/73
[51] **Int. Cl.²** **E02B 17/00;** E02D 5/72
[58] **Field of Search** .... 61/46.5, 53, 46; 114/.5 D; 175/9; 37/73

[56] **References Cited**
UNITED STATES PATENTS

2,941,369	6/1960	Quirin	61/46.5
3,183,676	5/1965	LeTourneau	61/46.5
3,256,537	6/1966	Clark	9/8
3,453,830	7/1969	Mitchell, Jr.	61/46.5
3,500,783	3/1970	Johnson, Jr. et al	114/.5 D

Primary Examiner—Jacob Shapiro
Attorney, Agent, or Firm—Torres & Berryhill

[57] **ABSTRACT**

An offshore drilling unit of the self-elevating type having a floatable hull and a plurality of legs movable from a raised position to a lowered position. Apparatus is attached to the legs for altering the response of the unit to movement of the body of water in which the unit is deployed. Apparatus for altering such response may comprise plating on the sides of the lower portions of the legs, longitudinal fins attached to the lower leg portions, flexible bags carried in the lower portions of the legs or any combination of such apparatus. The unit may also comprise a mass carried within the legs and mounted for longitudinal movement with respect thereto for altering the response of the drilling unit to water body movements. The movable mass may be used in conjunction with any of the other apparatus.

2 Claims, 9 Drawing Figures

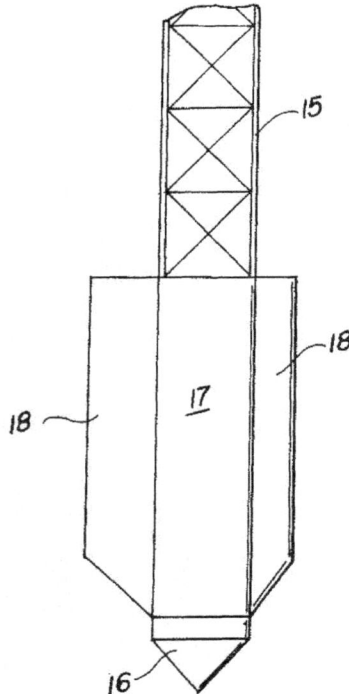

<u>Diagram 26</u>:
ETA's motion-tuning leg concept

2. <u>Going on location with high roll and pitch motions on the hull</u>: Diagram 26 shows the second feature proposed to improve going on location feature, again extracted from the patent.

Sea conditions could delay going on location and risk leg damage as legs hit seabed with horizontal loadings. The proposed solution provided three tuning measures for the floating jackup as it attempted to go on location:

(a) Plating in the sides of the lower lengths of the legs,

(b) Addition of drag flaps that could be extended on lower lengths of each leg, and

(c) Tanks or inflatable containers inside each leg truss which could be raised or lowered by winches on the hull to increase leg drag and inertia in the water and reduce hull rolling motions.

> <u>Patent</u>: *"Means for Altering Motion Response of Offshore Drilling Units," US patent no. 3,916,633, inventors: Peter M. Lovie & Edwin L. Lowery, filed August 24, 1973, granted November 4, 1975.*

Neither of these two features was ever used in any of ETA's designs.

Avoiding Leg Removal In Ocean Tows: A Novelty But A Third Patent Never Used

While we thought out ways to strengthen the leg design to avoid— or at least mitigate— the amount of leg to be removed in ocean tows, at one point we brainstormed a mechanical way to avoid any leg removal.

This was the third wild idea shown here in Diagram 27 which we patented but never used.

3   <u>Folding legs for ocean moves</u>: In the early 1970s, ocean moves for jackups normally required removal of top sections of the legs to reduce loadings on the legs at the hull level as the jackup hull rolled and pitched during the ocean move. Removal also kept the jackup floating stably and prevented overturning from its top-heavy leg configuration.

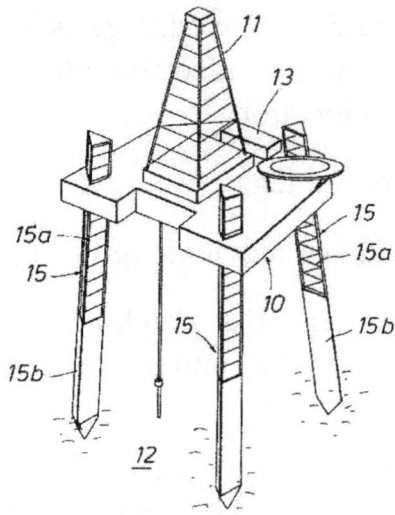

FIG. 1

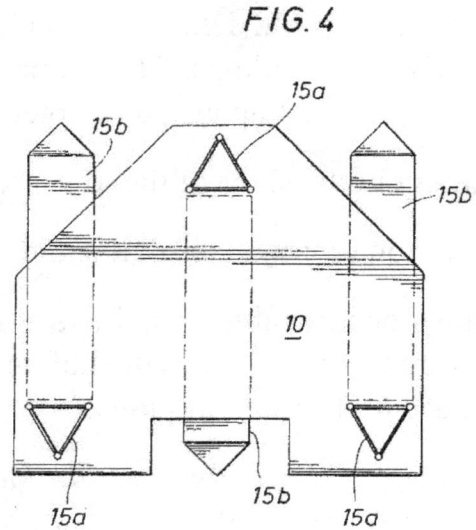

FIG. 4

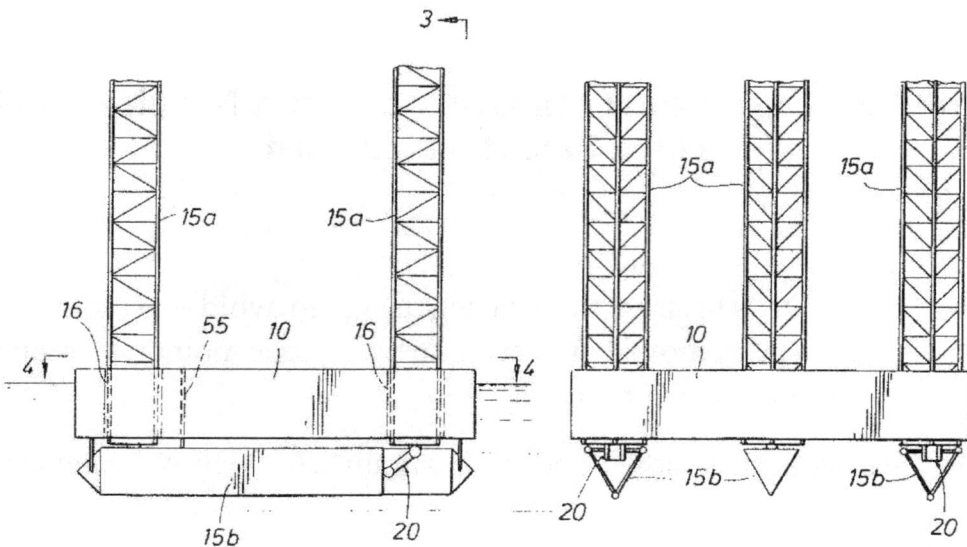

FIG. 2

FIG. 3

Diagram 27:
ETA's folding leg concept for ocean moves

Leg removal was slow and costly; the leg sections had to be transported with the jackup and then installed on arrival at the work location. The solution was to make the legs hinged, so that the lower portion of each leg folded under the hull for ocean moves.

There would be no need for cranes and frames for alignment and welding because sections of leg were removed and replaced on arriving at destination! The leg protruding above the hull was much reduced, so there would be no need to remove top sections and weld them back on after arrival. The hinged portion under the hull would then be unfolded and reconnected.

> Patent: *"Self-Elevating Offshore Platform with Folding Legs," U.S. patent no. 3,826,099, inventor: Peter M. Lovie, filed September 25, 1972, granted July 30, 1974*

As ETA designs developed, it was not necessary to use this idea. Parallel with this idea, we had developed leg designs to avoid leg removal entirely without any mechanical hinging. The ETA Robray 300 Class design would be able to do ocean moves with a 15 degree single amplitude criterion and the 425-ft.-long legs full up. The ETA Europe Class design had 508-ft.-long legs and was anticipated to do mostly field moves in the North Sea, which it could do with its legs full up.

A Fourth Patented Concept Never Used, Though It Inspired Four Other Developers

ETA's fourth never used patent is illustrated in Diagram 28 and was:

4. <u>A jackup production unit</u> that could also function as a mobile marine drilling unit was comprised of a floatable base; a floatable platform; and a single vertical leg attached to the base, such that the platform could jack up and down the single central leg. The base structure would be able to store oil produced from wells drilled from the platform. Diagram 16 shows front page of the patent with the drawing of the concept.

> Patent: *"Mobile Marine Drilling Unit," U.S. patent no. 3,996,754, inventor: Edwin L. Lowery, filed December 14, 1973, granted December 14, 1976*

United States Patent [19]

Lowery

[11] **3,996,754**

[45] **Dec. 14, 1976**

[54] **MOBILE MARINE DRILLING UNIT**

[75] Inventor: **Edwin L. Lowery**, Houston, Tex.

[73] Assignee: **Engineering Technology Analysts, Inc.**, Houston, Tex.

[22] Filed: **Dec. 14, 1973**

[21] Appl. No.: **424,838**

[52] **U.S. Cl.** 61/92; 175/7; 61/101

[51] **Int. Cl.²** E02B 17/00

[58] **Field of Search** .... 61/46.5, 46; 175/7, 175/9; 114/.5 D

[56] **References Cited**

UNITED STATES PATENTS

2,248,051	7/1941	Armstrong	61/46.5
3,244,242	4/1966	Wolff	61/46.5
3,433,024	3/1969	Diamond et al.	61/46.5
3,456,447	7/1969	MacKintosh	61/46.5
3,474,749	10/1969	Williamson	61/46.5 X
3,793,840	2/1974	Mott	61/46.5
3,898,847	8/1975	Magnanini	61/46.5

FOREIGN PATENTS OR APPLICATIONS

| 991,247 | 5/1965 | United Kingdom | 61/46.5 |

OTHER PUBLICATIONS

Ocean Industry of Mar. 1973, p. 30.
World Oil of Nov. 1973, pp. 90–92.

Primary Examiner—Jacob Shapiro
Attorney, Agent, or Firm—Bill B. Berryhill

[57] **ABSTRACT**

A mobile marine drilling unit comprising: a floatable base; a floatable platform; and a vertical support leg attached to said base and extending upwardly through a well provided therefor in said platform. Said base and a major portion of said leg being submergible in a body of water for support on the floor thereof. The support leg and platform are provided with elevating mechanisms for elevating the platform above said body of water on said leg. In deploying the drilling unit, the unit is floated to a selected site with the base drawn up underneath the platform and the leg extending upwardly through the well. When the site is reached the base is submerged with ballast until it is supported on the water body floor. Then the platform is elevated above the water body by the elevating mechanisms. A derrick may be moved over the leg well and drilling in the water body floor conducted through the well and leg.

5 Claims, 6 Drawing Figures

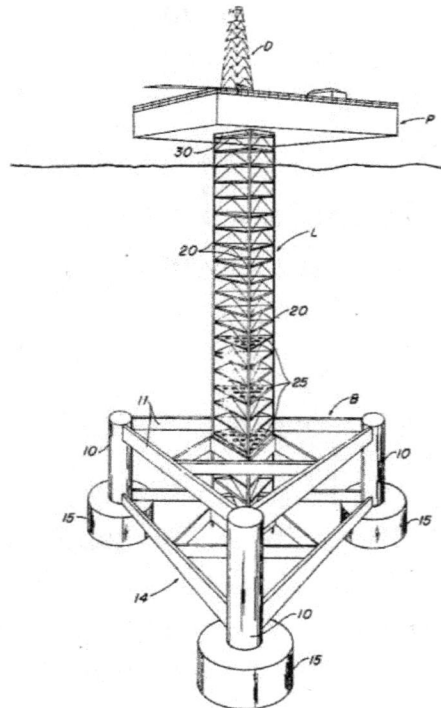

Diagram 28:
ETA patent on a jackup production system

In 1974, ETA thought about applying its jackup ideas to production_as well as drilling. This took the form of a patented concept for a jackup with one leg for use in up to 450-ft. water depths.

The idea was that this drilling and production platform could be towed out to location and in effect install itself. There would be no need for a derrick barge as commonly required for production platform installation. Later a derrick barge would be needed to install production packages.

The platform had a single triangular leg in the middle, founded on a triangular base that would be floated out, flooded, and set on the seabed. That triangular base looked like an early triangular Sedco semisubmersible.

The rectangular hull went around the single central triangular leg and jacked itself up and down. The buoyancy of the rectangular hull supported the central leg and triangular base as the base was lowered (jacked down) to the seabed.

We designed this concept to carry variables capacities that were larger than typical for jackups of the time but were similar to what was specified for the ETA Europe Class and the slant leg ETA Deepwater Jackup designs, shown earlier in Table E.

A Jackup With One leg: The ETA Mobile Monopod

The new concept was dubbed "The ETA Mobile Monopod" and publicized in articles in *Northern Offshore* and *Offshore* in 1974.

Diagram 29 shows the sequence of operations for mobilizing and installing the ETA Mobile Monopod, using a sequence of figures extracted from the patent. The general arrangement is shown in Diagrams 30, followed by the artist's impression in Picture 12.

The ETA Mobile Monopod stirred curiosity with its dual function for drilling and production. It had 20-24 slots for conductors down the central leg and oil storage in the lower hull.

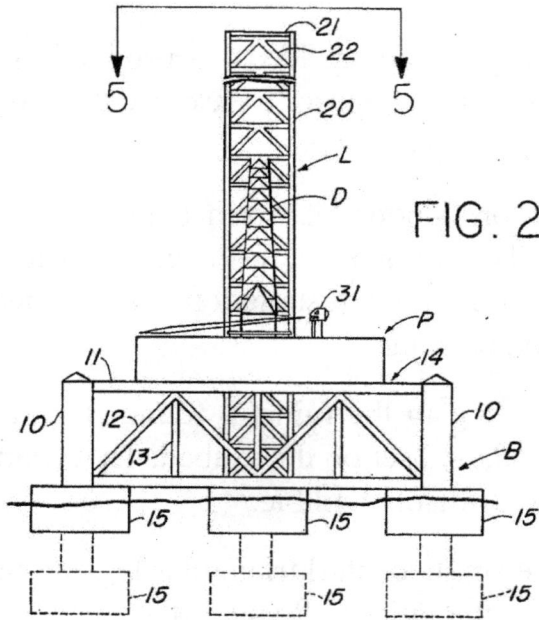

FIG. 2

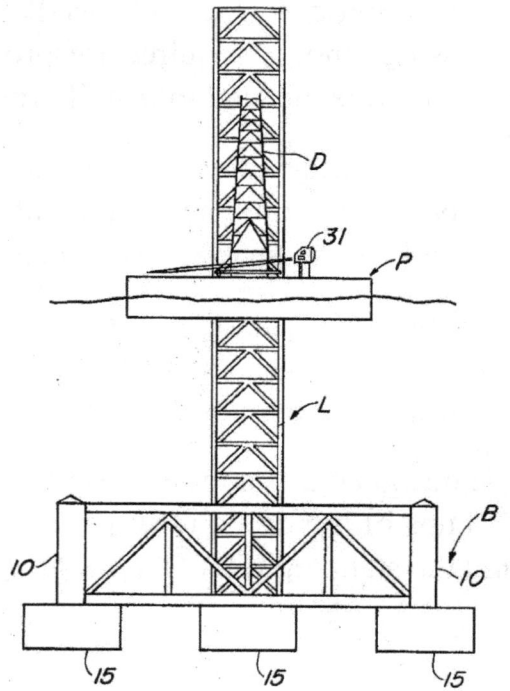

FIG. 3

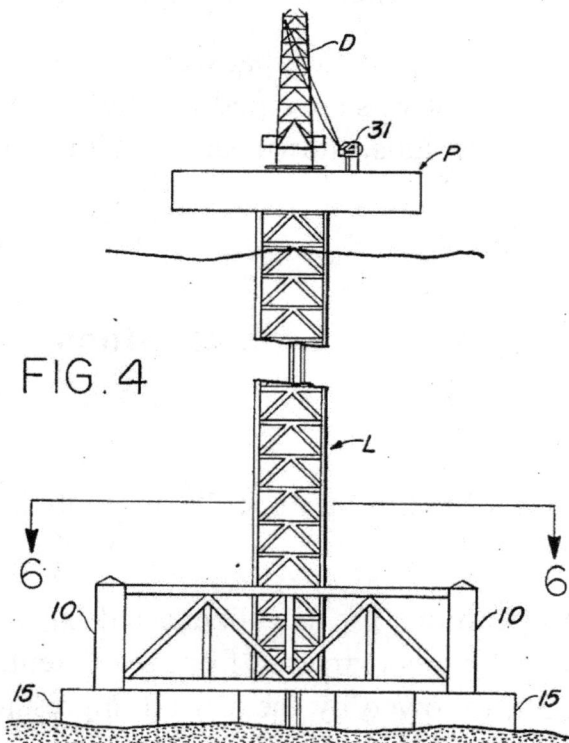

FIG. 4

Diagram 29:
Sequence in operation of the ETA Mobile Monopod: Fig. 2-arrival on location, Fig. 3- lower hull jacked down to seabed, Fig. 4-upper hull jacks out of the water tofinal operating position

Source: <u>U.S. Patent no. 3,966,754</u>, *filed December 14, 1973, awarded December 14, 1976*

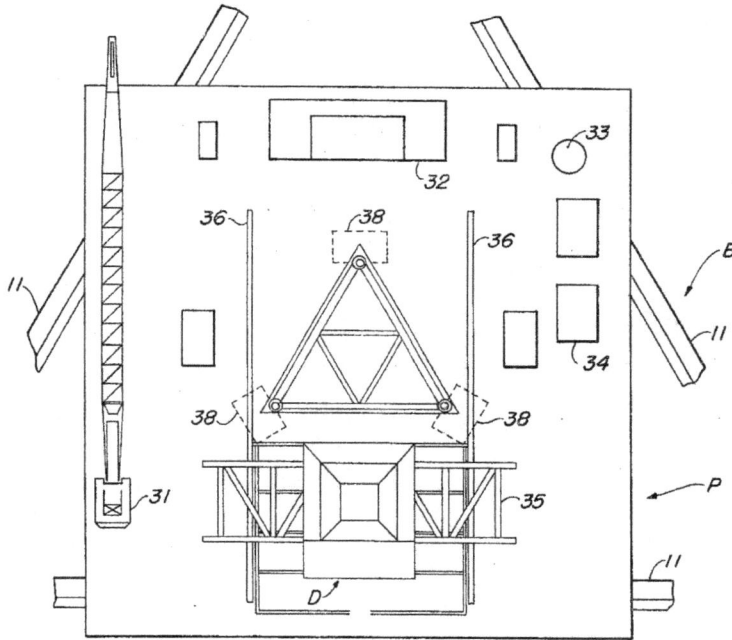

FIG. 5

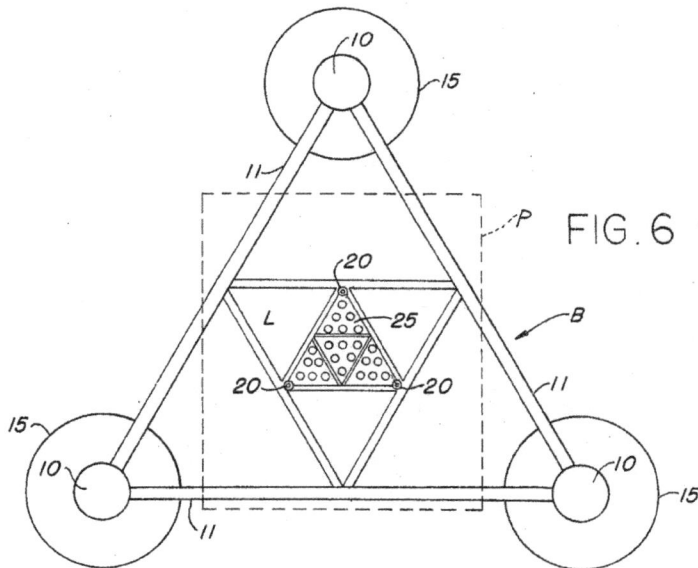

FIG. 6

Diagram 30:
General arrangement of ETA Mobile Monopod- Plan at deck of upper hull, plan of lower hull

About a year after ETA's Mobile Monopod was in the industry press, Texaco worked up an inquiry for use of a jackup production platform, apparently inspired by the ETA Mobile Monopod. The Offshore Company responded to the Request for Proposal, but nothing ever developed from that RFP.

The single-legged ETA Mobile Monopod, like the slant leg ETA Deepwater Jackup, was a bright idea, but it never got beyond the paper stage until 36 years after ETA had disappeared. The investment in engineering and patenting never paid off.

The illustrations from the patent and industry articles appear to have inspired other designers in future years to propose similar concepts…or perhaps ETA had just hit on an idea that had basic application in the industry.

Whatever the reason, during 1979-2013, four more designers in the U.S., France, and Singapore developed their version of the ETA Mobile Monopod concept.

These four designs are described and compared in the pages that follow, drawing on patents and technical papers.

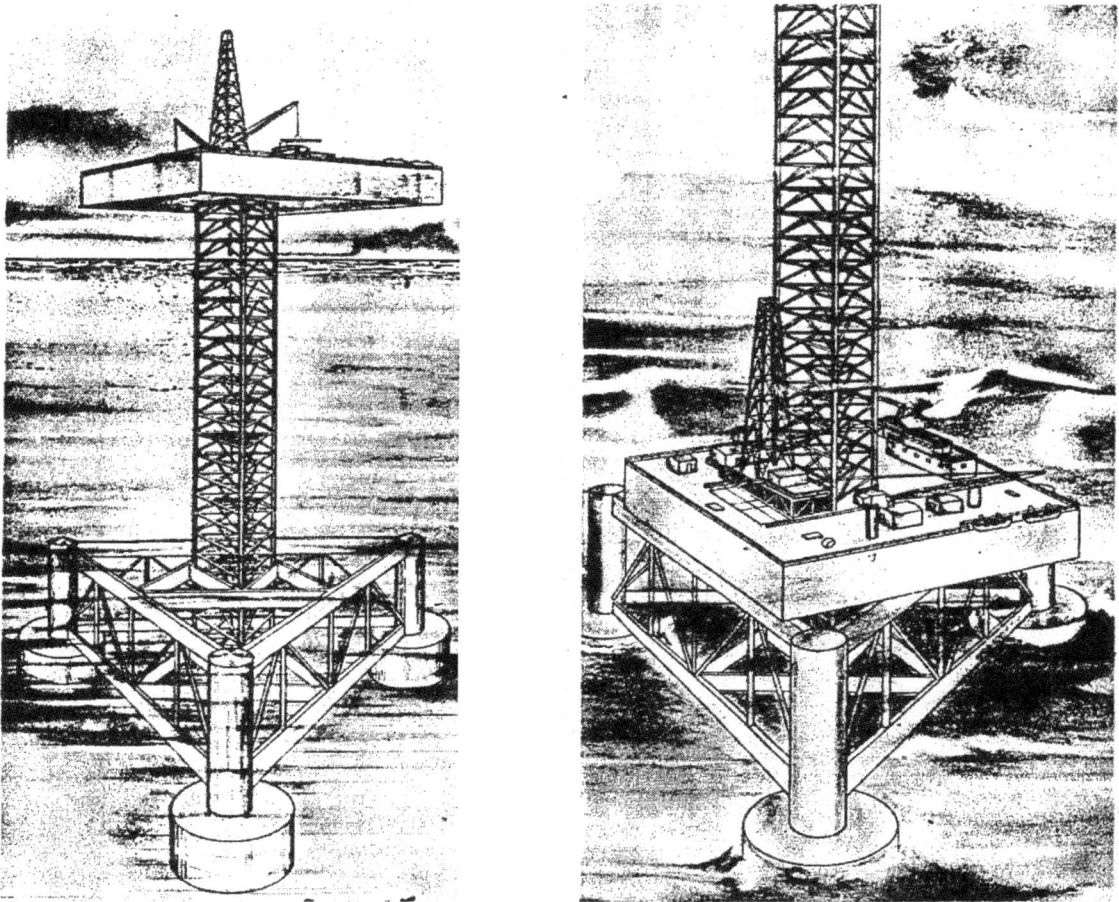

Picture 12:
The ETA Mobile Monopod while on location (left) and under tow (right)
Source: Articles in Offshore and Northern Offshore *in 1974*

The Second American Mobile Monopod (1979)

In 1979, the Offshore Company (TOC) in Houston, Texas filed a patent for a fairly similar design with a triangular leg and a rectangular barge hull which was adapted to use the proprietary TOC elevating system.

Diagram 31 is a composite of figures from TOC's patent in order to summarize their design.

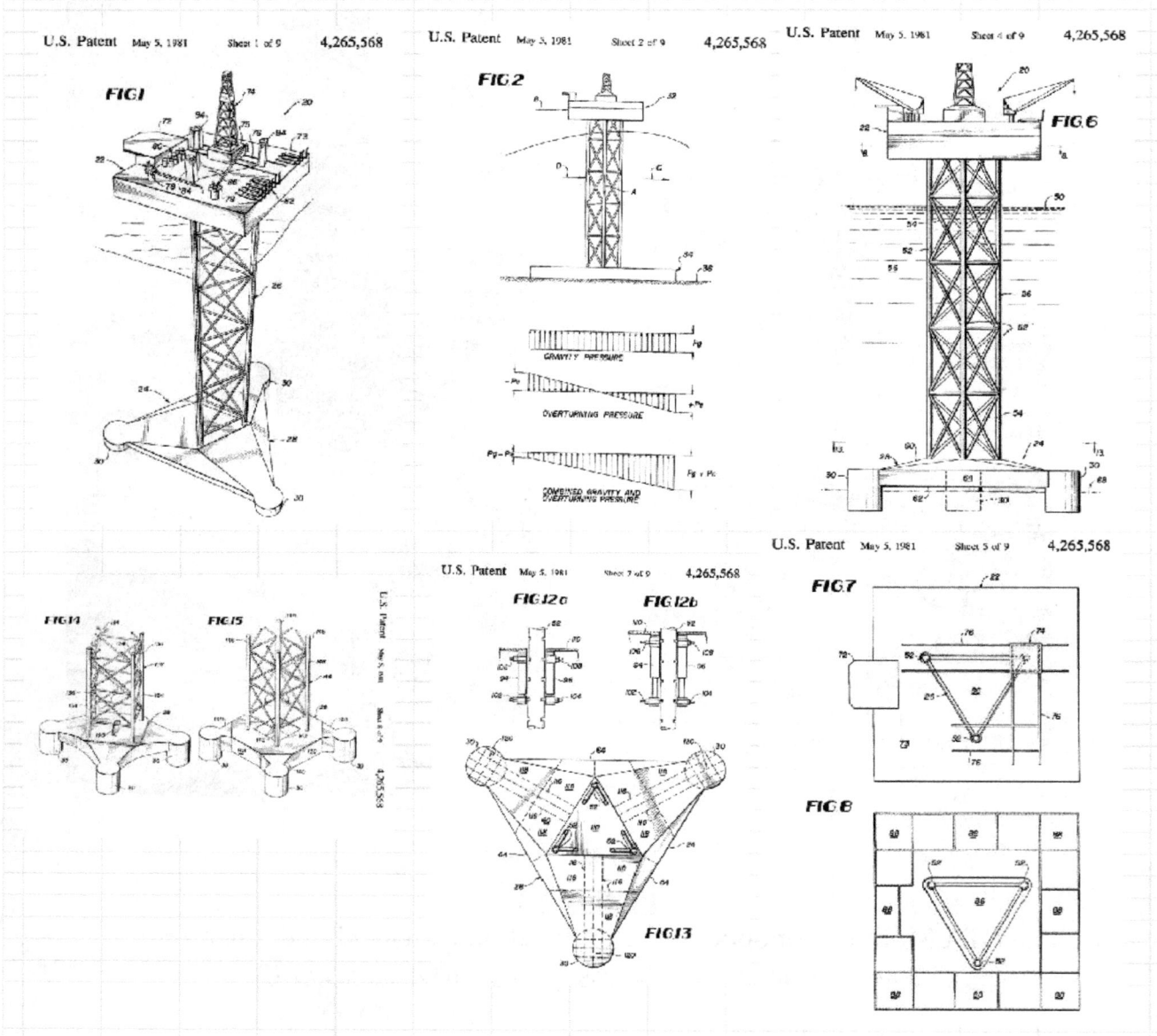

Diagram 31:

The Gravity Base Jackup Platform from The Offshore Company in Houston, Texas
Source: U.S. Patent no. 4,265,568, *filed August 6, 1979, awarded May 5, 1981*

TOC's design used a mat instead of the bottom structure with storage in ETA's Mobile Monopod. It was a triangular arrangement with circular footings at each corner, apparently to increase penetration and counteract the overturning effects of wind and wave. TOC made provisions for the leg to be square instead of triangular if desired. No documentation could be found to indicate what water depth had been in mind for the design.

The design did not appear to have been marketed; I was unable to uncover trade press articles about it.

The Third American Mobile Monopod (1984)

Diagram 32 is another composite of the figures in a patent. It shows the configuration of the Bethlehem 600, which is quite similar to its two predecessors. It had an approximately rectangular upper hull and a triangular central leg. Bethlehem Steel Corporation developed it at their jackup building yard in Beaumont, Texas.

The Bethlehem 600 made two serious advances: (1) capability of operation in up to 600 ft. of water worldwide, and (2) a means to adjust for the very real potential of the seabed being out of level—even sloping a degree or two, which could cause a significant operational drawbacks on the upper deck. The latter feature was very sensible; oil production locations do not always occur in perfectly flat seabeds as designers might desire!

Inventing a way to accommodate a sloping seabed location was significant—it was previously a weakness of the ETA design, somewhat less with the TOC design, and apparently ignored in the subsequent two designs of Mobile Monopods.

The ability to reach out to 600 ft. water depths was felt to be significant at the time, meant to expand the offshore area where a production jackup could be used. However, 1984 was a tough time to introduce the Bethlehem 600 in the oilpatch due to the collapse of oil prices and the deepening downturn. The potential for market expansion via the 600 ft. maximum water depth capability did not make a difference.

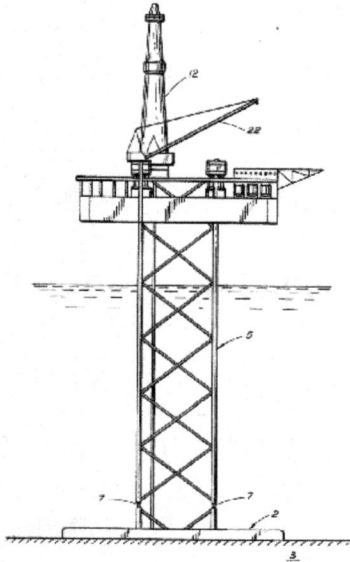

Fig. 15

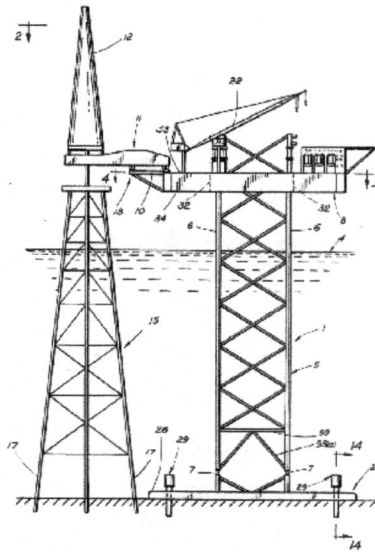

Fig. 1

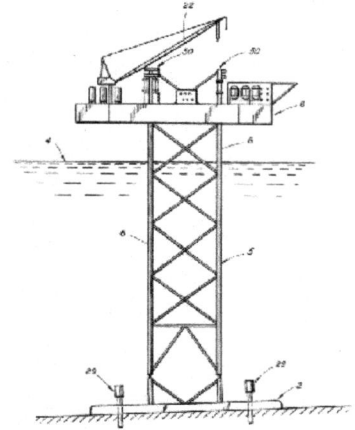

Fig. 6

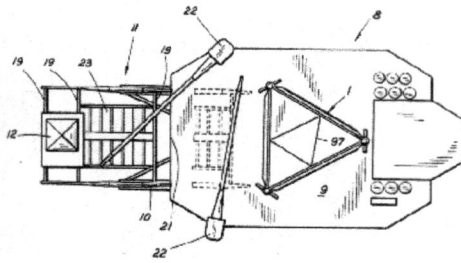

Fig. 2

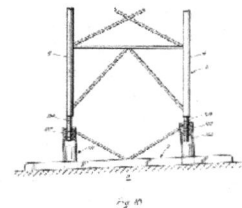

Diagram 32:

A "Mobile Offshore Jack-up Marine Platform Adjustable for Sloping Sea Floor" from Bethlehem Steel Corporation in Beaumont, Texas

Source: U.S. Patent no. 4,668,127, *filed April 22, 1986, awarded May 26, 1987*

Then A French Mobile Monopod (1991)

Seventeen years after ETA announced its ETA Mobile Monopod concept and fourteen years after CFEM delivered the two ETA Europe Class jackups, CFEM proposed their version of the Mobile Monopod concept in 1991 at the Offshore Technology Conference (OTC) in Houston.

Like its ETA predecessor, it used a rack and pinion elevating system with a single truss leg, and was designed for around 450 ft. of water. This version had a single square leg and was called the Jackup Monopile. Diagram 33 is a summary of OTC paper 6614, extracting figures from that paper to indicate the arrangement and operation of CFEM's Jackup Monopile concept.

The paper demonstrates serious engineering, apparently as part of a proposal for the development of the Balmoral field in the North Sea.

Because they borrowed heavily from previous concepts, CFEM's conclusion in the paper that they had "developed a new concept for drilling and production based upon a Jackup Monopile Platform with a gravity base and a self-elevating production deck" was not exactly accurate!

ETA had beaten them to their North Sea jackup innovations once again!

Construction At Last! A Chinese Mobile Monopod (2013)

Almost four decades after ETA created its ETA Mobile Monopod concept, published articles on it, and was awarded a U.S. patent on it, offshore drilling industry veterans Brian Chang (ex-Promet) and Peter Nimmo (ex-Bethlehem and Baker Marine Engineers in the U.S. Gulf of Mexico) proposed a similar single-leg jackup concept for drilling and production). Calm Oceans Pte. Ltd. of Singapore proposed this latest Mobile Monopod.

OTC 6614

The Jackup Monopile: A Production Drilling Platform

C. Perol, CFEM Offshore Engineering; S. Rodgers, Eiffel U.K.; and P. Maniere, CFEM Offshore Engineering

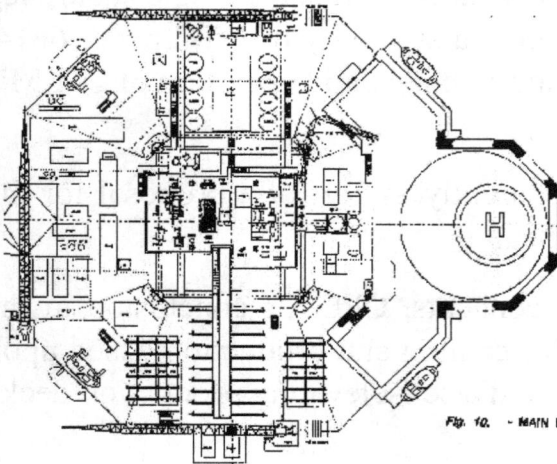

Fig. 10. - MAIN DECK LAYOUT -

Fig 1. ARTIST VIEW

CONCLUSION

CFEM Offshore Engineering has developed a new concept for drilling and production based upon a Jack-Up Monopile Platform with a gravity base and a self elevating production deck.

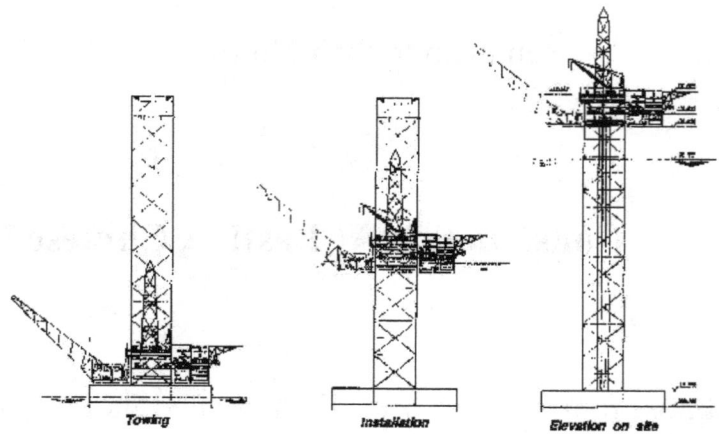

Towing   Installation   Elevation on site

Fig. 15. - INSTALLATION PROCEDURE -

Diagram 33:
CFEM's version of a Mobile Monopod
Source: Offshore Technology Conference, paper 6614, 1991

Their version was called the Mono Column Platform (MCP). The MCP is designed for up to 500 ft. of water (versus ETA's 450 ft.) and has a square leg instead of a triangular leg.

An MCP is now one of the many jackups under construction on speculation in China and is waiting for a buyer. Picture 13 shows it in an artist's impression and then during construction.

Picture 13:
A Chinese Mobile Monopod under construction, circa 2015
Source: www.bcholdings.com

Funnily enough, in the 1970s another meaning for the abbreviation MCP was a "male chauvinist pig." If in ETA we had thought to instead name our Mobile Monopod an MCP, our marketing program might have had a field day! But thanks to the passage of time and distance, an MCP over in China could be socially acceptable in today's world!

On April 14, 2014, Nantong Blue Island Offshore Co., Ltd. and Calm Oceans Pte. Ltd. held a steel-cutting ceremony for their *Calm Ocean 101* MCP jackup platform at the Nantong Blue Island shipyard in Qidong, China.

Storage And Offloading Ignored With Mobile Monopods

Neither ETA nor any of the other subsequent designers of Mobile Monopods addressed the arrangement for exporting oil and gas production and whether some form of oil storage would be necessary if no nearby pipeline was available. The ETA Mobile Monopod did provide for limited storage in its lower hull to cater for production testing.

Storage and offloading could be ignored in most field developments in the U.S. Gulf of Mexico, where there is a broad network of pipelines to receive oil and gas production. In contrast, in most other parts of the world, the export of production might not have been as easy, demanding use of a storage tanker to be stationed nearby, perhaps moored to a CALM Buoy for regular offloading to export tankers. In the descriptions of the five Mobile Monopod designs cited in Table G, none touch on this concern.

Export is a situation that Calm Ocean Pte. Ltd. as designer and builder will have to deal with if they are successful in finding a buyer for the *CALM Ocean 101*.

Comparison Of The Five Mobile Monopods Of 1974-2013

The ETA concept was considered for a North Sea field development but did not progress past paper. The TOC Gravity Base Jackup Platform and the Bethlehem 600 design did not get past the paper stage either. Seventeen years after ETA's Mobile Monopod, the CFEM concept was considered for a specific field development in the North Sea but again did not go past the paper stage.

Calm Ocean's MCP progressed further and has gone into construction on a speculative basis and at the time of writing, in 2018, more than three years after cutting first steel, it does not appear to have a buyer,.

Table G summarizes a comparison of the features of ETA's original Mobile Monopod concept in 1974 and the four subsequent similar concepts.

Designer:	Engineering Technology Analysts, Inc. (ETA)	The Offshore Company (TOC)	Bethlehem Steel Corporation	CFEM Offshore Engineering S.A.	Calm Oceans Pte. Ltd
Design publicly announced:	1974	apparently not announced	1984	1991	2013
Design name:	ETA Mobile Monopod		Bethlehem 600	Jackup Monopile	Mono Column Platform (MCP)
Location of design company:	Houston, Texas	Houston, Texas	Beaumont, Texas	France	Singapore
Max. water depth, ft.:	450	???	600	492	500
Variables capacity, kips:	8,500	???	???	4,184	14,313
Leg configuration:	Triangular truss	Triangular truss	Triangular truss	Square truss	Square truss
Upper hull:	Rectangular barge	Rectangular barge	"Octagonal" barge	Octagonal barge	Rectangular barge
Lower hull:	Truss, 3 large round footings	Triangular mat	Rectangular	Rectangular, rounded corners	Rectangular
Jacking system:	Rack and pinion	TOC pins & jacks, hydraulic	Bethlehem pins & jacks, hydraulic	Rack and pinion	Rack and pinion
No. of well conductor slots:	24	???	???	24	9
Progress with design:	Concept first developed, North Sea in mind	Allowed for square or triangular leg, triangular mat with footings at corners	Concept similar to TOC but allowed for out of level seabed	Concept engineered for North Sea	"CO 101" Prototype building on spec in China, started construction in 2014
Primary information source:	U.S. Patent no. 3,966,754, filed 14Dec73, awarded 14Dec76	U.S. Patent no. 4,265,568, filed 6Aug79, awarded 5May81	U.S. Patent no. 4,668,127, filed 22Apr86, awarded 26May87	OTC paper 6614, 1991, 6 pages, 4 tables, 16 figures	website: www.bcholdings.com

Table G:
Comparison of the five Mobile Monopods of 1974-2013

How did other companies come to latch on to ETA's Mobile Monopod concept? TOC and Bethlehem, which were both active in the Houston community, would know the ETA design and have a sound feel for operator reactions.

CFEM had known ETA well from their building the two ETA Europe Class jackups and would have seen the published articles about ETA Mobile Monopod. Similarly, the president of Baker Marine Engineers (Peter Nimmo) would have recalled the ETA patent and design from his days at BME during 1977-1981.

It remains interesting how four different designers, followed the original ETA Mobile Monopod concept, each in their own way. And after forty-three years, only one of these structures has actually been built!

Any rights to the ETA's Mobile Monopod design and patent had long ago expired, similar to the TOC and Bethlehem patents.

Once again, ETA's ideas in 1974 were ahead of their time.

PART V

FROM PAPER TO "IRON":

CONSTRUCTION AND HISTORY OF JACKUPS BUILT FROM ETA'S DESIGNS

After ETA did its part in executing its design contracts, jackups were built using these designs and in 2018, people now ask about the history with the jackups delivered 35-43 years ago. Part V gives that history as best can be assembled.

In chronological order by design contracts, there were three groups of jackups that were built worldwide from ETA designs:

1. The ETA Robray 300 Class jackups built in two shipyards in the Far East: at least nine;

2. ETA Europe Class jackups: two built at CFEM in France for Dyvi Drilling A/S. They were the biggest in the world at that time and were designed to record demanding standards. In 1981, the Singapore delivery of a third jackup (*Zapata Scotian*) drew on the ETA Europe Class design. *Zapata Scotian* was intended for operation in the shallower but harsh waters of Eastern Canada;

3. Two series of substantially smaller workover and production drilling jackups. Pool Company commissioned ETA to design jackup platforms to satisfy their specific operations and accommodate their drilling and service equipment and systems. A total of ten were built in Denmark, Singapore, South Africa, and the U.S.

Groups 1 and 2 employed the features covered in ETA's patent on jackup leg design, U.S. 3,967,457, described earlier. The third group was quite different. With the designs owned by Pool, Group 3 used some new ideas developed by ETA but more simply with less scope than in Groups 1 and 2.

ETA Robray 300 Class Jackup Design: Nine Built

Current industry data sources (e.g. Rigzone) show Robin Shipyard in Singapore ultimately delivered five ETA Robray 300 Class jackups: one to Brazil, two to China, and two to India.

Picture 14:
The *Norbe I,* an ETA Robray 300 Class jackup, delivered in 1979 for drilling, then converted 1991-1992 to an accommodation unit for North Sea
Source: Society of Petroleum Engineers

Four additional ETA Robray 300 Class jackups were delivered from Hitachi's Innoshima shipyard in Japan. Since 1976 onwards, deliveries had a typical service life of 25-30 years, so it would be reasonable to expect most of these jackups would have already been retired or been lost in accidents by 2018. In 1985, the *Nanhai III* was lost in a blowout, but the eight others are still in existence: five working, two stacked, and one converted. The *Norbe I,* shown in Picture 14 is now an accommodation unit called the *COSL Rigmar,* located in the North Sea.

A Chinese Reincarnation Of The ETA Robray 300 Class Jackups

Akin to how Robray had wanted one of ETA's people in the shipyard during construction, Dyvi Drilling likewise wanted one of ETA's engineers in France at CFEM. Ramesh Maini, a bright energetic member of the engineering team in ETA, had been in the middle of all the analyses and design calculations with me in 1974. He went to work at the shipyard in France as ETA's liaison. Later he went to work for Dyvi Drilling to assist rig movers in operations with the *Dyvi Beta* and *Dyvi Gamma*.

Four decades later, Ramesh Maini and Rao Guntur now lead Zentech, an engineering company. They steadily built it up with a large staff and developed a remarkable jackup engineering practice. Persevering through often difficult times, their business has culminated in the creation of a deepwater jackup design for 400 ft. of water for Far East service. The first of these was delivered in 2016 and currently sits in Huangpu Shipyard in Guangzhou, China, waiting for work, as shown in Picture 15 on the left.

In a historical sense, it marks a remarkable juxtaposition for China's past and present. China has gone from importing ETA Robray 300 Class jackups from Robin Loh in Singapore and Hitachi in Japan to building dozens, even hundreds of jackups and hoping to export them to the rest of the world!

The other side to this change is that China has gone hog wild, taking business risks in building jackups for entities building on speculation with no contract.

In this regard, the Chinese have taken far larger risks than drilling contractors in ETA's day or by U.S. and Norwegian investors a few years later. .

In 2017, dozens of new jackups are complete, ready for work, and parked at multiple shipyard docks. In Picture 15, the fuzzy satellite image on the right, indicates nine jackups sitting at a dock, and the shadows of their legs visible on the sea signal how large they are. It seems that extremely favorable Chinese financing packages in the face of emptying shipyards helped create a supply of about 292 jackups more than demand (Bassoe, November 2016).

The Zentech R550D design (image on left of Picture 15) can be said to symbolize a reincarnation of the ETA Robray 300 Class jackup again intended for Far East service: same hull depth, same leg chord configuration, same leg plan, but of course with up-to-date systems and equipment and forty years of fundamental Houston engineering progress of these ETA alumni.

Picture 15:
China's forty-year turnaround
Source: Waiman Kwan

Pioneering The Construction Of Jackup Legs With Cast Steel Joints

Cast steel joints were a key feature in the ETA jackup leg design, as shown in the diagrams and design drawings in Part III, "Design Philosophies," and included in U.S. patent no. 3,967,457. When we first used cast steel joints in the ETA Robray 300 Class jackup design I thought it made good engineering sense, and nobody could prove me wrong! Hitachi built two of these jackups in Japan for Robray and later two more for China. Robray Offshore Drilling Co. and Robin Shipyard in Singapore had common ownership via Robin Loh. Shortly after Robray Offshore Drilling Company received the design package of drawings and specifications from ETA, we learned that Robin Shipyard was expected to build ETA Robray 300 jackups and they had started to investigate the supply sources of steel castings, in Thailand and Malaysia.

Deliveries of batches of cast steel joints from these foundries were found to be outside allowable dimensional tolerances and on cutting apart were found to have porosity problems. This was a surprise to us in ETA, as we felt the dimensional and material tolerances were not that difficult and there were well-experienced foundries in the region, e.g. our repeated telexes recommended foundries in Japan capable of delivering what was needed for such key components in jackup legs as joint pieces. In fact, five years later, three major Japanese rig builders offered to supply cast steel jackup leg nodes.

We never did understand Robin Shipyard's foundry selection when other reliable sources were available. For generations foundries have cast manhole covers to dimensional tolerances that fit snugly into their frames, installed by the millions in streets everywhere! Foundries had been delivering cast steel piping components for many years that were not unlike the ETA cast steel nodes in dimension, design, and steel quality. To demonstrate the acceptability and design of these cast steel joints, ETA engineers took Robin Shipyard's chief engineer to see ABS in New York. We well understood how building jackups was a daunting task and a completely new activity for Robin Shipyard. We wanted to show how cast steel nodes could satisfy stringent regulatory requirements and how construction using them was not that tough and might possibly be simplified!

Perhaps procurement of cast steel joints was too much to deal with in the startup or the shipyard management simply did not like building from a jackup design that had not been built before. It was well known that building from a new design was not the same as building another (say) LeTourneau or Bethlehem unit from very well-tried plans and fabrication procedures! Robray had chosen the new ETA design to outperform these other jackups. ETA never learned why Robin Shipyard made the decision not to use cast steel joints in the jackups built in their yard.

Not using cast steel joints meant a critical change in the leg design, such as one of the following:

(a) Keeping the original bracing member wall thicknesses sizes and reducing environmental and tow criteria to compensate for increased joint stresses;
or

(b) Beefing up the tubulars and chords from the original design;
or

(c) Using some form of thicker wall stubs at the joints and then the original design of bracing members between the stubs.

It's not certain which of these three choices was made for the jackups built by Robin Shipyard, but comments heard from engineers from that era indicate that thicker wall stubs (c) probably were the choice.

Piecing the history together today, it appears that delays at Robin Shipyard may have contributed to Robray Offshore Drilling's forfeiture of a drilling contract for their first jackup. Consequently, Robray insisted that their jackups be built at a more established yard, Hitachi in Japan. Suffice it to say there seemed to have been a rocky start to jackup building at Robin Shipyard. That is borne out in a 1978 memo from the Val Meadows, chairman and managing director of FELS, to Larry Baker Sr., president of Baker Marine in Texas (reproduced in Diagram 34). Meadows comments about Robin Shipyard, "As we have always known it was the shipyard that screwed things up."

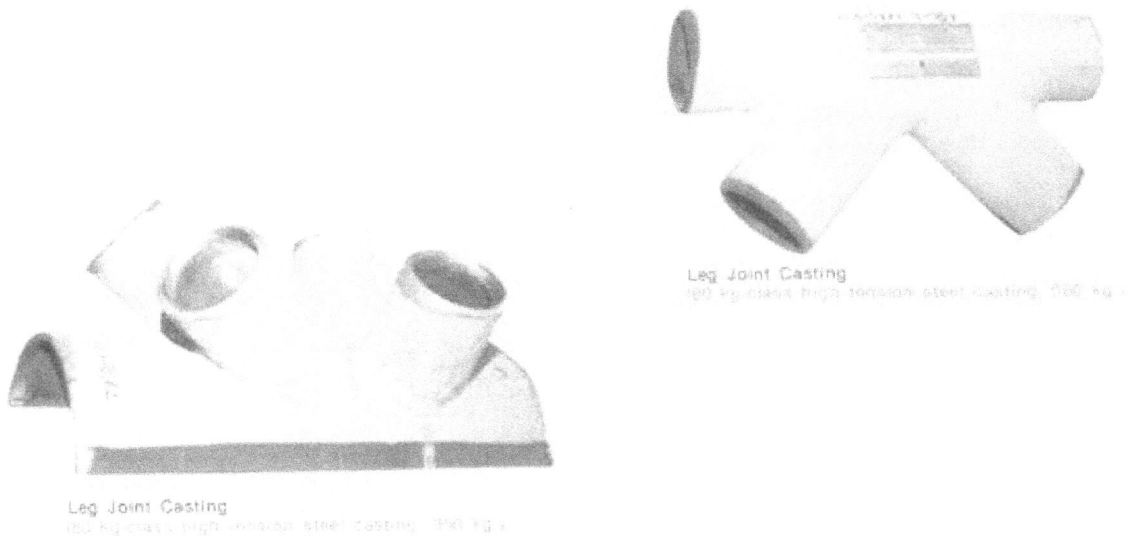

Picture 16:
ETA cast steel joints for leg chord and K brace connections, manufactured in Japan for Hitachi for the ETA Robray 300 Class jackups
Source: Canam Construction

Kobray Offshore Only

FAR EAST-LEVINGSTON SHIPBUILDING LIMITED
(INCORPORATED IN THE REPUBLIC OF SINGAPORE)

TELEX: RS 21513	31 SHIPYARD ROAD,	POSTAL ADDRESS:
CABLE: "FESHIPIND"	JURONG TOWN	P.O. BOX 6,
TEL. NOS: 652144 (8 LINES)	SINGAPORE 22.	JURONG TOWN POST OFFICE
		SINGAPORE 22.

P.H. MEADOWS
CHAIRMAN/MANAGING DIRECTOR

MEMORANDUM

TO: Mr. Larry A. Baker, Sr
        Mr. Peter Lovie
        Mrs. Emily Tschang

FROM: P.H. Meadows

DATE: November 7, 1978

-- Please see page 2, item "Robin Gives Licence to Japanese Yard", Singapore Investment News.

This development dispels doubts about ETA design! As we have always known, it was the shipyard that screwed things up. But it means that BMC/Lovie/FELS need not stand away from its identity with the ETA design.

BMC will doubtless look into Robin's rights to sell design to Hitachi.

P.H. MEADOWS

PHM:fl

Diagram 34:
Memo from Far East Levingston Shipyard to Baker Marine
Corporation, dated November 7, 1978
Source: Lovie & Co. files

His statement supports the reputation of ETA designs. Robin Shipyard blamed its own yard problems on ETA and its designs. That talk negatively impacted ETA in the relatively small offshore drilling community and probably held back ETA's progress. Picture 16 shows what the famous cast steel jackup joints for ETA Robray 300 Class jackups looked like in real life. These are castings from a foundry in Japan for use in ETA Robray 300 class jackups built at Hitachi.

Despite the scuttlebutt in Singapore, ETA designs kept passing regulatory vetting, and subsequent years of operation counteracted adverse criticism from the late 1970s.

Robin Shipyard managed to build jackups of ETA Robray 300 Class design in Singapore but without using steel castings, delivering a total of five jackups to buyers in Brazil, China and India during 1976-1982, when other shipyards in Singapore delivered even larger numbers of MODUs.

Picture 17 shows leg fabrication with cast steel joints for the two ETA Europe Class jackups. The image is extracted from an OTC paper published in 1978.

A total of six jackups were built by two shipyards, using ETA's jackup leg designs with cast steel joints as Table H shows.

Picture 18 shows a stack of cast steel joints at CFEM's yard in Dunkerque, including both the leg chord and the K brace castings, ready for incorporation into the leg sections.

With the passage of time, it has become clear how the cast steel joints may have been great for structural efficiency and understandable to structural engineers. However, nobody (including ETA as a designer) attempted a thorough investigation into effects on fabrication procedures, which would have been beneficial.

I have been unable to find reports on the maintenance and structural performance of the cast steel leg joints but can conclude that a 41-year-old jackup with cast steel joints that is still working in the North Sea at least shows the robust reliability of the feature!

Picture 17:
ETA cast steel jackup leg joints at the leg chord to bracing member connection on the ETA Europe Class jackups, fabrication in CFEM's yard in Dunkerque, France
Source: <u>Offshore Technology Conference</u>, *paper 3244, 1978*

Like Hitachi, CFEM had substantial fabrication, procurement, project management, and engineering resources, and was an experienced fabricator of offshore structures at the time. They had access to multiple suitably skilled foundries that could supply the necessary castings.

One lesson learned was how cast steel joints really needed to be used with an established fabricator.

Shipyard	Jackup name	Original Owner	Current Owner	Delivered	Status at end of 2016
	ETA Europe Class:				
CFEM	*Dyvi Beta*	Dyvi Drilling A/S	Petrobalic	1976	Working: converted to MOPU, Poland
CFEM	*Dyvi Gamma*	Dyvi Drilling A/S	- - -	1977	Lost in accident, North Sea, August 21, 1990
	ETA Robray 300 Class:				
Hitachi	*Ednastar*	Robray Off. Drlg	North Korea	1976	Renamed *Yu Son*, stacked in China
Hitachi	*Ednarina*	Robray Off. Drlg	CNOOC	1977	Renamed *Bohai IV*, drilling in Bohai Bay
Hitachi	*Nanhai III*	PRC Marine	- - -	1980	Lost in blowout, Southeast Asia, 1985
Hitachi	*Nanhai IV*	PRC Marine	COSL	1980	Drilling in Bohai Bay

Table H:
Jackups built using ETA's design of cast steel joints in their legs

Picture 18:
Leg K brace connections: Cast steel joints at CFEM's yard during fabrication of legs for ETA Europe Class jackups. Note the larger leg chord joint piece at upper right.
Source: Waiman Kwan

Three Leading Japanese Offshore Fabricators Offer Supply Of Cast Steel Nodes For Jackup Legs (1978-1980)

After vetting by classification societies (ABS and DNV) and government regulators in the North Sea, the first jackups with cast steel nodes were delivered in 1976 and 1977. Afterward, leading Japanese fabricators with a strong track record of building for the offshore market came forward with their proposals for supplying this new component for use in building jackups. They were three big names each well able to perform: Hitachi, Kawasaki, and NKK.

It was a logical move for Hitachi, which had employed cast steel nodes in the delivery of *Ednastar* and *Ednarina* to Robray Offshore Drilling Co.

They could now add this new kind of steel casting to things they routinely supplied to the marine business, such as steel castings for crane hooks, valve bodies, and anchors. Hitachi cited features of their cast steel nodes for jackup legs to include:

- "Low temperature impact performance, even suitable for use in the North Sea;"

- "Tensile properties to match high strength steel plate used in offshore structures;"

- "Welding operations at structural pipe connections are greatly reduced;"

- "Cracks which had been caused in profusion at concentration of welds at pipe connections in the past can be prevented."

All these features are sound and along the lines ETA had discussed. Hitachi illustrated their brochure with images of the ETA Robray 300 Class design.

I don't know just what Hitachi's marketing thrust was. Use of cast steel nodes was a fundamental choice early in the development of a design and one that the end customer would want to be part of, i.e. the drilling contractor that might own and operate the jackup for many years. To overcome natural past preferences for what had been successful before, this product would also require basic marketing to jackup designers at an early stage.

When Hitachi's brochure was printed (excerpt shown in Diagram 35) and the marketing of their nodes started, it must have been an uphill battle, with the leading jackup designers at that time being designer-builders that had their own designs and were reluctant to change! The originators and enthusiasts for ETA's cast steel node were no longer around.

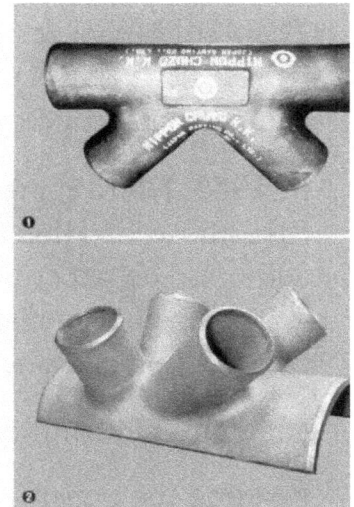

<u>Diagram 35</u>:
Excerpt from Hitachi's 1978 technical brochure on cast steel nodes for jackup legs

There was little doubt Hitachi could supply the cast steel nodes to whatever high standards were needed, but sales and marketing were tough.

A year later, in 1979, Kawasaki came out with their version of a proposal to supply cast steel nodes of similar high standards. Diagram 36 shows their Prenode product line in an excerpt from their technical brochure. Unlike Hitachi, they had not built jackups with cast steel nodes, but they took the trouble to build a section of leg with cast steel nodes to demonstrate their capabilities and learn from that experience. Kawasaki's technical data and rationales were generally similar to those of Hitachi.

The third Japanese offshore fabricator offering to provide cast steel nodes was Nippon Kokan K.K. (NKK). NKK had set up a cooperation agreement with ETA in 1977 that had been inherited by Baker Marine Engineering (BME).

In 1980, NKK came forward with both a technical proposal via a technical brochure (like those of Hitachi and Kawasaki) and a short technical paper by a team of six authors. The paper described their testing of the cast steel concept using photo elastic and strain gage methods on a full scale set of jackup leg joints: one set with conventional construction and one set with cast steel nodes.

Diagram 37 identifies and shows an excerpt of the paper.

Reading the paper in 2018, I think it is clear that the paper doesn't fully describe the joints being tested, probably due to lack of space in the journal in which the article is printed. It is not clear how rigorous the comparison was: whether the comparison is with a built-up, conventional joint with thicker stubs at the joints or whether the joint had a normal pipe thickness.

The cast steel nodes show about four times longer fatigue life in some of the numbers and some beneficial features but it was not conclusive in reading this in 2018. It is likely that the published paper is a summary; perhaps a comprehensive report would be able to define the comparison more clearly.

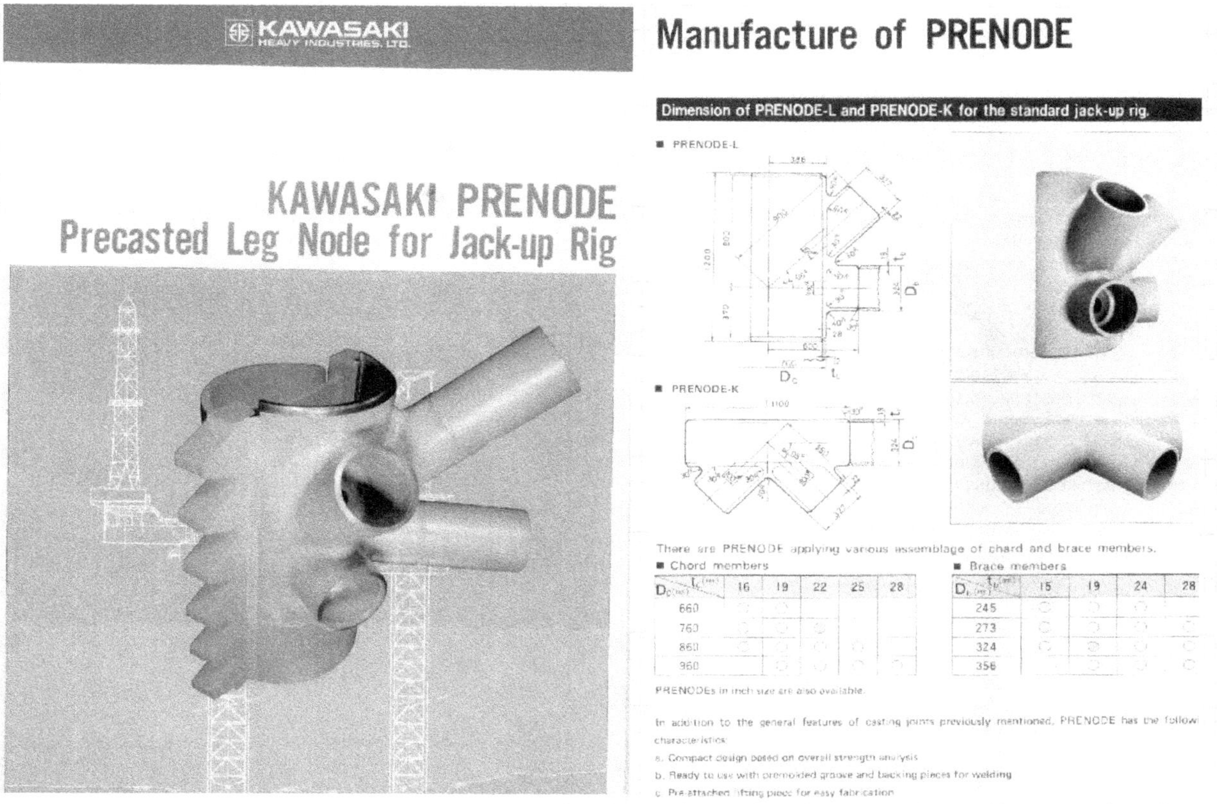

KAWASAKI HEAVY INDUSTRIES, LTD.

Manufacture of PRENODE

KAWASAKI PRENODE
Precasted Leg Node for Jack-up Rig

Dimension of PRENODE-L and PRENODE-K for the standard jack-up rig.

■ PRENODE-L

■ PRENODE-K

There are PRENODE applying various assemblage of chord and brace members.

■ Chord members

D_c (mm)	16	19	22	25	28
660					
760					
860					
960					

■ Brace members

D_b (mm)	15	19	24	28
245				
273				
324				
356				

PRENODEs in inch size are also available.

In addition to the general features of casting joints previously mentioned, PRENODE has the following characteristics.

a. Compact design based on overall strength analysis
b. Ready to use with premolded groove and backing pieces for welding
c. Pre-attached lifting piece for easy fabrication

<u>Diagram 36</u>:
Excerpt from Kawasaki's 1979 technical brochure on cast steel nodes for jackup legs

When the article was published in 1980, NKK was building jackups of multiple designs that did not use cast steel nodes. Therefore, it is not clear what the underlying objectives of the study were and who had asked for it.

The practical matter of introducing this innovation had succeeded. The first jackups at Hitachi and CFEM had chosen to follow ETA's specifications with their drilling contractor customers.

Whether or not ETA had been on the right track, in 1980 no offshore drilling contractor clamored to use cast steel nodes on their jackups. The reality was that builders and owners just did not go for cast steel nodes post-ETA. There were no other jackups built with cast steel joints in the years that followed.

TECHNOLOGY

Japanese Study
Jackup Rig Leg Node Fatigue Testing

By Tsutomu Nakajima,
Mitsuo Konomi, Akira Kunitomi,
Takeshi Yano, Atuo Nagashima
and Kozo Asano
NKK (Nippon Kokan K.K.)

With the continued emphasis on offshore drilling and oceanic development under rigorous conditions, the jackup has become an important rig. But its obvious advantages, three or four retractable legs providing excellent stability in waves, pose problems. A review by NKK indicates that on approximately half of the jackup rigs in service, trouble of one type or another is likely during their service life because of the engineering of the legs.

Structural characteristics of the jackup rig are as follows:

- Legs must bear all vertical and horizontal forces acting upon them (even though the deck may not be directly subject to wave forces);
- Rigidity of the joint between the legs and lifting gears is unclear;
- Strength of the leg portions which sit on the sea bed is unclear;
- Legs are unstable during towing;
- Leg lengths vary with water depths; and
- The low rigidity of the legs renders the rig susceptible to the repeating loading of waves, etc.

As is widely known, all offshore

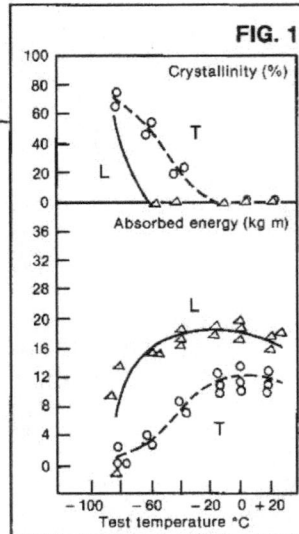

mensionally accurate and practically defectless nodes being successfully developed.

Built-up Node
This node consists of a half chord and four braces. For the steel plate, NKK-manufactured 80 kg/sq mm-class high tensile steel NK-HITEN 80 B was used.

Charpy, impact tests (2mm V-notch) have revealed that this steel is so cryogenically tough it

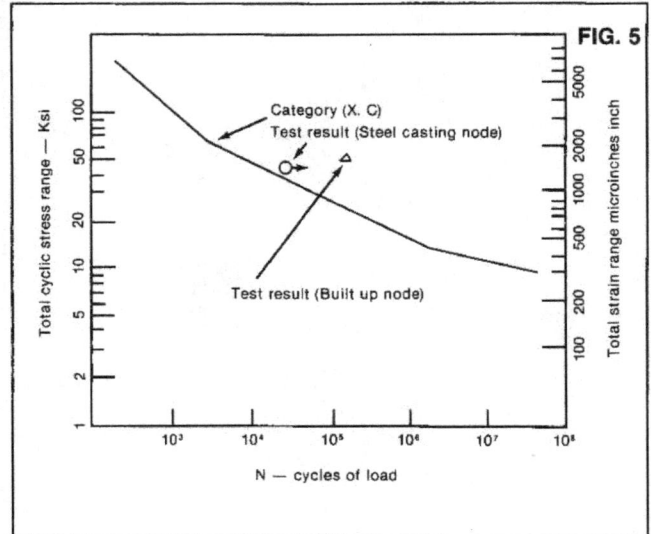

tion tested the butt joints.

Casting Node
When thick-walled, high tensile steel pipes are used, as in jackup rig legs, the welding of leg nodes is difficult and impractical. Therefore, one-piece nodes of casting steel were made in an attempt to improve strength and cut welding operations. As in the case of built-up nodes, an 80-kg/sq mm-class, high-tensile cast steel is used for

gonal braces and chords, an outline stress distribution in a node had to be evaluated. The stress distribution obtained in the photoelastic coating process was then used as a reference to bond a strain gage (Figs. 2 and 3) and determine the stress in a joint, particularly the stress distribution in the hot spot.

Also note in Figs. 2 and 3, relative deformations of the horizontal brace and of a chord were mea-

Diagram 37:
Excerpt from NKK's technical paper on cast steel nodes for jackup legs
Source: Tsutomu Nakajima et al, "Jackup Rig Leg Node Fatigue Testing," Oil & Gas Digest, *pp. 10-11, September 1980*

A question that remains unanswered is how much of a difference the cast steel joints really made in structural efficiency and resistance to fatigue cracking over the lives of actual jackups. It would be eye-opening to see a rigorous estimate of the true extent of potential weight savings, fabrication economies, and operating history.

ETA's Leg Chord Design: A Pioneering Advance That Became Widely Used

Jackup designs have elements that are more demanding than others, although they have to function as an integrated system. For example, the hull is relatively straightforward to lay out and design. The importance shifts to the arrangement of equipment and systems. As water depths became greater and design loads more severe, the legs become particularly important and greater amounts of analyses become necessary. The leg chord becomes particularly significant because it is both a critical strength component, as well a weight element for motions and stability. In addition, it must work efficiently with the rack and pinion elevating system that has become the elevating system of choice.

ETA felt that its leg chord made leg fabrication simpler. While we did not realize it at the time, the mid-1970s were a time of great progress in jackup design and the start of the all-time jackup building boom cycle of 1978-1986.

When ETA designed its deepwater jackups, the industry precedent consisted of two configurations for the leg chord design to accommodate a rack and pinion elevating system.

The first was a roughly triangular single rack design of leg chord, which dated to 1956, when R.G. LeTourneau delivered the first ever rack and pinion jackup in its yard in Vicksburg, Mississippi. This leg chord design was successfully used with LeTourneau jackups ever since.

In the late 1960s Levingston Shipbuilding in Orange, Texas had developed a more recent leg chord design, using a tubular leg chord with two opposed racks mounted on the outside of the leg chords and offset from the center. This leg chord design was designed to work with the National Supply elevating system and was first used in their jackup for Marlin Drilling, delivered in 1969, using a square truss leg in contrast to ETA's triangular truss legs.

ETA had also planned on using the National Supply elevating system and needed to have a two-rack opposed arrangement to be able to work with the National Supply system. ETA felt better structural efficiency could be achieved by having these racks on center in the legs. This arrangement would avoid the eccentric vertical loads present in the Levingston design while still using a tubular leg chord concept for lower wave and current loadings.

The Levingston chord was 40 inches in diameter, perhaps for fabrication convenience. (This probably accounted for the need to have offset racks so that the frame for the National Supply rack and pinion elevating system would be practical in its design.)

ETA chose to use a leg chord diameter that was much smaller at 30 inches which solved the elevating system constraint, reduced wave and current loads, and made for a much more efficient structure.

There were horizontal components of the loads imposed on the leg chord from the pinions, and we accommodated those by having a diametral plate that separated the two racks and yet balanced that horizontal load effect.

The ETA leg chord design was included in U.S. patent 3,967,457 discussed earlier and illustrated in the related diagrams.

As the jackup industry expanded, there was a total of four commonly used leg chord designs for rack and pinion elevating systems. In addition to the LeTourneau, Levingston, and ETA designs, the Friede & Goldman design came along later and gained a lot of support in 1978 onwards. It is a tubular chord with an opposed on-center rack. However, the rack is doubled to make the cross-sectional area of the leg chord comparable to the other tubular designs while keeping the diameter even smaller (15 inches) to reduce wave and current loadings.

As 1978 started, ETA had gone out of business and Friede & Goldman moved to capitalize on the huge demand for jackups. All four pioneering leg chord designs are now shown in Diagram 38.

The information in Diagram 38 is extracted from the report by the Society of Naval Architects & Marine Engineers (SNAME) on "Guidelines for Site Specific Assessment of Mobile Jackups," Technical & Research Bulletin 5-5A, Revision 3, August 2008, 366 pages. This broad industry study had its first report in 1994, led initially by Shell. It followed much of the DNV-led work in the North Sea during the 1980s that had started with the ETA jackups. The listing of working panel members is quite remarkable; it covers the majority of jackup industry technical leaders of the day. It is thus a valuable, credible, and detailed industry document.

A useful follow-up document is the ISO Standard of 2012, which covers jackup analysis and design and includes much of the data on leg chords. These standards illustrate a remarkable advance on design standards from the ETA's days!

All four of the leg chord pioneers in Diagram 38 were in the United States. It is no surprise that all four were located on the U.S. Gulf Coast, where offshore drilling was pioneered years earlier and continued to evolve.

By examining the particulars on jackup leg chord designs in Figures 8.1.1 and later at the end of the SNAME study, then correlating with *Rigzone* data for jackup design sources during 1970-1986, it becomes possible to discern how different leg chord designs gained acceptance, including those of ETA.

The ETA leg chord design had been the first of the tubular chords with double central racks, the configuration shown in ETA's patent. It made good engineering sense and was used in a number of jackups, even long after ETA had gone out of business!

Just counting jackups delivered by 1986, a total of 39 jackups used the leg chord design introduced by ETA in 1973 and covered in the U.S. patent no. 3,967,457. With 12 of these being ETA-designed jackups, it meant there were 27 other jackups which used the ETA leg chord design. That upstart ETA had done something right!

The ETA leg chord design had become much more popular than ETA's cast steel joints!

The four pioneering leg chord designs in Diagram 38 are identified again in Table I below with the year of first design and other designers that later employed their leg chord design.

Table I shows how in the years after ETA had gone, four other designer-builders chose to use ETA's pioneering leg chord arrangement in their own designs. CFEM and Hitachi already had experience fabricating jackup legs for ETA designs and employed the ETA leg chord design in their own jackup designs. Baker Marine and Modec also used it in their designs.

Two U.S. Gulf Coast Independent Design Firms

E T A
(out of business 1977)

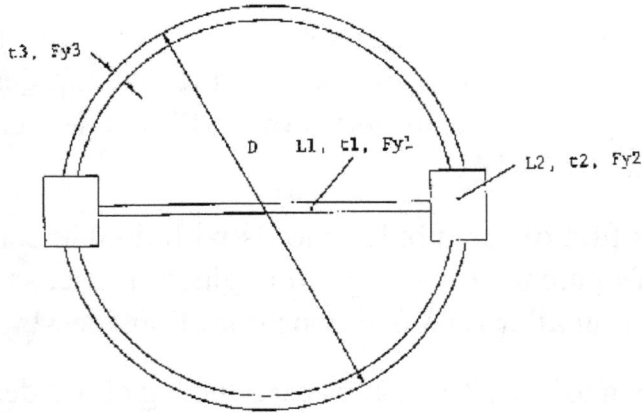

Friede & Goldman
(acquired by CCCC of China in 2010)

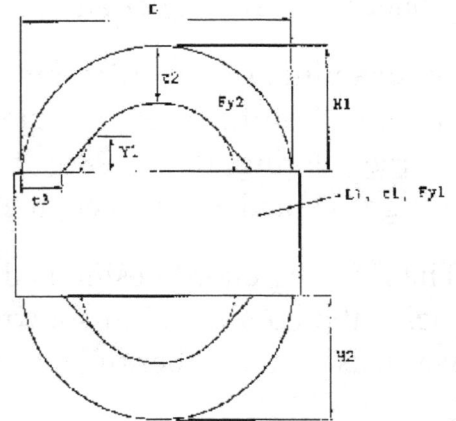

Two U.S. Gulf Coast Pioneering Jackup Designer-Builders

Levingston
(out of business 1983)

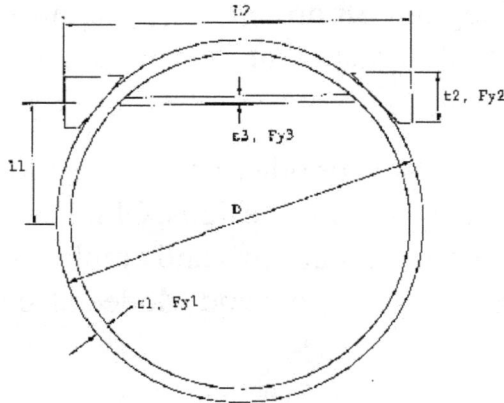

LeTourneau
(acquired by KeppeFELS 2016)

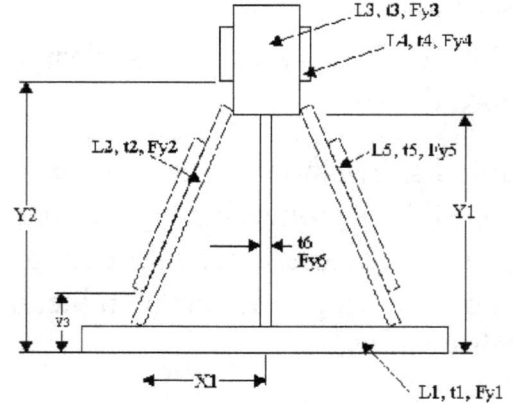

Diagram 38:
Rack and pinion jackups: Leg chord designs from the four pioneers
Source: Extracted from Figures C8.1.1-4 in "Guidelines for Site Specific Assessment of Mobile Jackups." <u>Society of Naval Architects & Marine Engineers</u>, *Technical & Research Bulletin 5-5A, Revision 3, August 2008, 366 pages*

Background	Leg chord design pioneer	Location	First design	Chord dia., in.	Subsequent users of leg chord design
U.S. Gulf Coast Independent Design Firms	Engineering Technology Analysts (ETA)	Houston, TX	1973	25	Baker Marine, CFEM, Hitachi, Modec, Robin
	Friede & Goldman	New Orleans, LA	1978	15	MSC, Technip
U.S. Gulf Coast Designer-Builders	R.G. LeTourneau	Vicksburg, MS	1956	Not applicable	Many LeTourneau designs, 200+ built
	Levingston Shipbuilding	Orange, TX	1968	40	Hitachi, Modec, Robco

Table I:
Rack and pinion jackups: The four leg chord design pioneers and subsequent use of their designs by others

Turning to the details of their leg chord designs, Table J gives dimensions and wall thicknesses, all extracted from the SNAME study.

From a design standpoint, the leg chord design did not have to be identical for each jackup design; it could be varied quite easily in dimensions, as Table J now indicates. ETA's philosophy in the ETA Robray 300 and Europe Class jackups was to keep the leg chord wall thickness constant but to vary the thickness of the internal diametral plate. The chord diameter and the bay height were the same for both designs.

Loadings on the ETA Europe Class leg chords dictated that they had to be much heavier than the ETA Robray 300 Class. Thus the diametral plate thickness varied anywhere from 38-140 mm thick, i.e. at some places near the top of the legs, the leg chord had a slab that was rack thickness all the way across. That was not the case with the lighter Robray 300; its diametral plate varied from only 10-32 mm. thick, and the rack was slightly thinner at 127 millimeters.

				For key to dimensions see Diagram 34: Rack and pinion jackups - leg chord designs for from the pioneers									
Design finalized	Designer/ Builder	Design	Est. No. Delivered 1976-1986	L1	t1	L2	t2	D	t3	Fy1	Fy2	Fy3	Bay height
				Dimensions are in millimeters, Yield Stresses in Mpa									
1978	Baker Marine	JU-300-CAN (*Zapata Scotian*)	1	991	127	0	0	914	44-48	690	0	690	5,532
1979-1980	CFEM	T2001 (Hitachi redesign)	1	960	18	121	140	960	26-52	690	690	690	4,050-4,100
		T2005	6	650	20	108	140	800	28-42	700	685	650-700	5,050
		T2600	1	650	20	120	140	800	33-58	700	700	700	6,000
1973	ETA	ETA Robray 300 Class	9	627	10-32	127	127	762	22	690	690	690	5,486
1974		ETA Europe Class	2	627	38-140	140	140	762	22	690	690	690	5,486
1979-1980	Hitachi			900	18-20	100	127	900	32-50	690	690	690	5,160
		K1025/31/32	3										
		K1026 (*Neddrill 4*)	1	950	18-20	100	127	950	32-42	690	690	690	4,360
		K1056/7	2	1000	28	130	178	1,000	47-64	690	730	690	4,600
1978-1980	Modec	200	2	450	15-27	102	127	559	27	490	690	490	5,486
		300	10	450	25-115	102	127	559	34-40	490	690	490	5,486
		400 (Trident 9)	1	690	20-35	102	127	800	30-35	490	690	490	6,200
		Estimated total built:	**39**										

Table J:

Characteristics of the opposed two rack on center: ETA leg chord design, used by ETA and other designer-builders in jackups delivered during 1976-1986
Source: Extracted from Figures C8.1.1-4 in "Guidelines for Site Specific Assessment of Mobile Jackups." Society of Naval Architects & Marine Engineers, *Technical & Research Bulletin 5-5A, Revision 3, August 2008, 366 pages*

Along with the dimensional data, Table J identifies the builders and numbers of jackups built with the ETA leg chord design.

Among the other jackup builders using the ETA leg chord design:

Baker Marine kept the slab of rack thickness all the way up and down the leg length and varied the tubular wall thickness;

CFEM chose to keep the diametral plate at a constant thickness and varied the tubular plate thickness;

Hitachi and Modec chose to vary both the thickness of the diametral plate, as well as the tubular wall thickness.

Bay heights varied from one designer-builder to the next. Thus there was a lot of flexibility in the ETA leg chord design.

Status Today Of The ETA Europe Class Jackups: The *Dyvi Beta* and *Dyvi Gamma*

Picture 19 shows *Dyvi Beta* operating in its original drilling mode when owned and managed by Dyvi Drilling A/S of Oslo, Norway.

In 2015, *Dyvi Beta* was converted to become the *Petrobaltic Beta*, working for LOTOS Petrobaltic as a mobile offshore production unit (MOPU) in Polish waters and is still in existence in 2018, shown in Picture 20. .

The second ETA Europe class jackup, also owned and managed by Dyvi Drilling A/S was the *Dyvi Gamma*. It started life as a drilling unit, as shown in Picture 21, in 1977 leaving the shipyard en route to its first location at Ekofisk in the Norwegian sector of the North Sea, where it would work for Phillips.

Later *Dyvi Gamma* was converted to an accommodation unit (Picture 22). It was lost when a line broke under tow in a storm in the North Sea on August 21, 1990. The crew of 49 ended up in the water, but all were rescued.

Picture 19:
The ETA Europe Class jackup design: The *Dyvi Beta* in 1977
Source: Sea Education Association

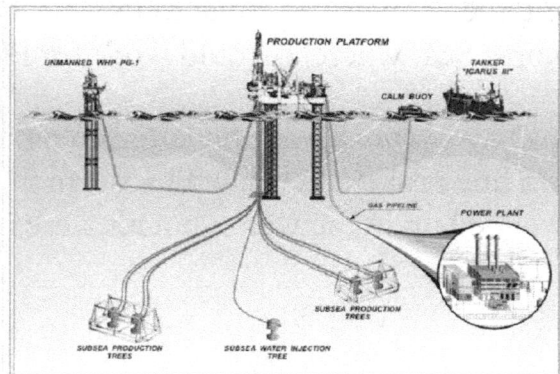

Picture 20:
Still going strong in 2017: The *Dyvi Beta,* now *Petrobaltic Beta*, in Block B3 in
Baltic Sea as a MOPU
Source: LOTOS Petrobaltic

<u>Picture 21</u>:
ETA Europe Class jackup design: The *Dyvi Gamma* leaving the shipyard in 1977
Source: www.upi.com

Picture 22:
The *Dyvi Gamma* after conversion to an accommodation unit
Source: www.Lovie.com

Fabrication Of *Dyvi Beta* And *Dyvi Gamma* By CFEM

Unlike in Singapore, in Northern Europe the climate is much colder, and frequent rain makes fabrication outdoors more difficult. CFEM chose to fabricate sections of the legs for *Dyvi Beta* and *Dyvi Gamma* indoors in a fabrication hall, as shown in Picture 23. At the bottom left, it is possible to see the cast steel joints at the leg chords with the stubs awaiting connection to bracing members.

CFEM reported on the fabrication of *Dyvi Beta* and *Dyvi Gamma* in OTC paper 3244 with key dimensions now given in meters and weights given in metric tonnes rounded off to the nearest hundred. The paper starts with citing the approvals granted by Norwegian regulators and from DNV as a classification society.

In 1971-1972, ETA made floating stability analyses of the *Pentagone 82* semisub-

Picture 23:
Indoor fabrication of leg sections for the ETA Europe Class jackups at CFEM's yard
Source: Waiman Kwan

mersible. The client was LeTourneau, which was building at their yard Brownsville, Texas. The name of the *Pentagone* came from the originator of the design, Andre Rey-Grange of Forex Neptune in France because of its <u>five</u> columns. Understandably, in 1978 CFEM chose to follow the lead of their countryman and rename their two three-legged ETA Europe Class jackups "Trigones." Their OTC paper no. 3244 was "Trigone Jackup Drilling Rig."

As an aside in the small world of MODU designers, in 2007, the Oilfield Energy Center in Houston inducted Andre Rey-Grange into their Hall of Fame as an industry pioneer, in large part because of his Pentagone design, which offered unusually low levels of motion. During 1973-1975, I was fortunate to get to know Andre Rey-Grange during talks about ETA jackup designs with Forex Neptune at their office in Montrouge, just outside of Paris. However I was not aware back then of his inspiring the Trigone moniker given to the ETA Europe Class jackups!

OTC paper no. 3244 is 8 pages long with 10 pictures and 1 diagram. Two of these pictures are extracted and included here: one of the steel castings shown earlier in Picture 18 and one in the left of Diagram 39, illustrating the leg installation.

The right of Diagram 39 reproduces a plan indicating the leg design, essentially a repeat of parts of earlier diagrams derived from ETA's U.S. patent no. 3,967,457.

The image at the left of Diagram 39 shows how the legs were assembled on the ETA Europe Class jackups (a.k.a. Trigones!). A crane on the deck of the jackup set each leg section of about 36 ft. length, with a cabin around the leg chords to protect welders, preheat equipment, and ultrasonic testing units from the elements. Leg erection consumed a reported 400,000 welding rods to make the leg connections. It must have taken quite a number of cycles of elevating and lowering additional leg sections to reach the final leg length of 508 ft.

OTC 3244

TRIGONE JACK-UP DRILLING RIG

by Gilbert Laplante and Rene Clauw,
Compagnie Francaise D'Entreprises
Metalliques

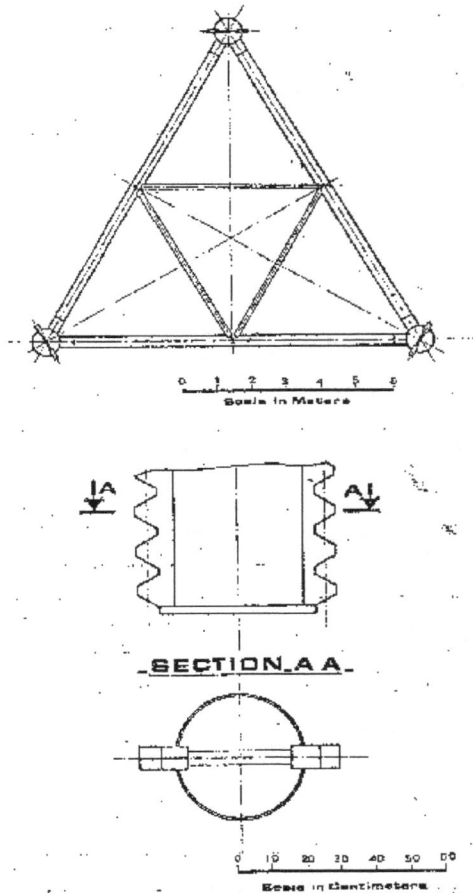

FIG. 3 - TYPICAL SECTION OF CORNER CHORDS AND TRUSS OF LEGS.

Diagram 39:
Fabrication of *Dyvi Beta* and *Dyvi Gamma* as reported by CFEM
Source: Offshore Technology Conference, paper 3244, 1978

In the paper, no mention is made of ETA or ETA Europe Class! At the offshore industry's leading conference in the world, this paper remained silent on design responsibility for the precedent-setting pair of the world's largest ever jackups in 1977. The Houston audience noticed the omission, but most blew it off. *That's the French for you, over there in the North Sea. They cannot stand a little Houston company upstaging them on a design for their home turf.* Now CFEM's behavior is ancient history, obviously water under the bridge, but it was indicative of the climate in 1977.

In our naiveté, ETA had felt our pioneering way would receive industry recognition, but such was the real world.

ETA Europe Class Design Helped Shape Future Design Practices

When Det Norske Veritas (DNV) classed these jackups, they vouched for its safety to Norway's regulator, the Norwegian Petroleum Directorate (NPD). It was a time when the industry felt record water depths for jackups and the North Sea's harsh environment required verification of valid environmental loading theories. DNV assigned a top team to examine the design's validity in the face of actual loadings.

In 1977, *Dyvi Gamma* experienced its first winter in the North Sea and was instrumented with strain gauges to measure the environmental loads it was subjected to while on location at the Ekofisk complex (see Picture 23, where this big jackup looks tiny in relation to the Ekofisk complex). The instrumentation of *Dyvi Gamma* was part of the study that DNV and Statoil performed starting in November 1977.

Weather was suitably rough to provide good data! Ramesh Maini, Dyvi's rig mover and former engineer on the ETA Europe Class design team, reported, "I was onboard the rig when all the measurements were taken because we had 45 ft. seas continuously for three weeks and we couldn't move the rig."

The forces and overturning moments developed in *Dyvi Gamma* that stormy November were measured at the locations indicated in Diagram 40 and reported during 1978 in Veritas report no. 78-326.

This study was followed by an expanded study effort in 1980-1983, again led by DNV.

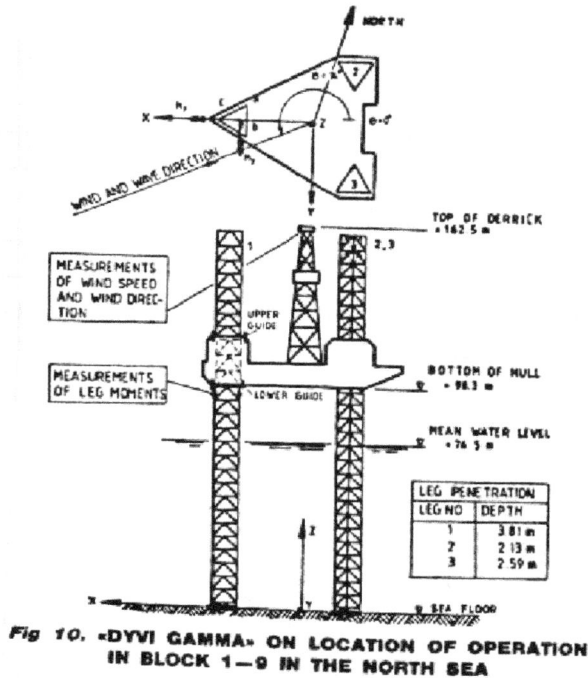

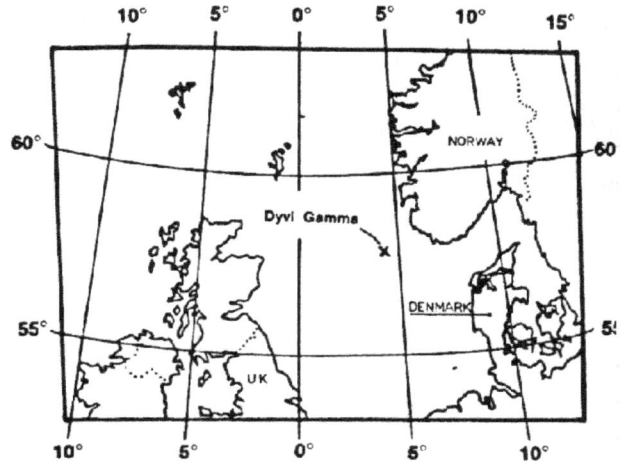

Fig 10. «DYVI GAMMA» ON LOCATION OF OPERATION IN BLOCK 1—9 IN THE NORTH SEA

Fig 9. GEOGRAPHIC LOCATION OF «DYVI GAMMA» DURING MEASURING (EKOFISK)

<u>Diagram 40</u>:

Dyvi Gamma participates in jackup storm response measurements program in the North Sea starting November 1977

Source: "The Jackup Drilling Platform," <u>Gulf Publishing</u>, 1985, paper by Carl-Arne Carlsen et al, pp. 90-136

This time five operators (BP, Phillips, Shell, Statoil, Unocal) and two MODU designers and builders (Gotaverken and CFEM) supported the study, which considered the usual wind, wave and current effects, appropriate values of drag coefficient, etc. DNV realized that one loading effect might be conservative (safe) while another might offset it, causing an unsafe condition. How might these effects used in jackup design be combined for a sound assessment of the overall safety of the jackup?

Diagram 40 extracts an image from a paper published in 1985 that pulled together all the DNV study work from 1977-1985.

For the ETA Europe Class jackup designed in 1974, the summary data in Table K shows : (1) ETA had always been safe in using the Stokes Fifth Order wave theory for the full range of operating water depths, although the margin of safety went down from 1.3 to 1.1 as water depths increased from 70 to 110 meters, and (2) in line with industry practice of 1974, we had taken *static* loadings but had ignored the second order deflection *dynamic* loading effects. This caused the structural margin of safety to go from 0.93 to 0.77 over the range of 70 to 110 meters of water.

DNV claimed that in 1985, including that dynamic loading in combining everything, the margin of safety was 20% more than needed for 70 meters of water. This met what was necessary for 90 meters of water, but was 15% lower than what was necessary in 110 meters of water.

TABLE 2　　Typical example of implicite safety factor for simplified response analysis due to waveloading.

Analysis Assumption	Water depth (m)			
	70	90	110	130
A. Regular wave	1.5	1.5	1.5	1.5
B. Wave theory Stokes 5th Order	1.3	1.2	1.1	1.05
C. Drag Coefficient Cd=0.5	0.65	0.65	0.65	0.65
D. Static Analysis 1)	0.93	0.87	0.77	0.63
Total built in safety factor	1.2	1.0	0.85	0.65

1) 3 legged jack-up with lattice type legs

Table K:
Design guidance initiated from the *Dyvi Gamma* studies
Source: "The Jackup Drilling Platform," Gulf Publishing, 1985: paper by Carl Arne Arnesen et al on DNV led studies, pp. 90-136

Interpolating Table K to ETA's water depth ratings, it indicates that the ETA Europe Class design stacked up as follows:

Water depth, ft.:	250	300	350
1985 DNV safety rating:	1.06	0.99	0.87

After all these years of studies, it turned out that ETA had done a decent job of design after all!

The dynamic effect was nonlinear, so the second-order dynamic effect with these long legs at greater water depths made good sense when you think about it! It would be expected that the dynamic or second-order effect became much more significant in greater water depths, as Table K shows.

DNV commented:

> *Based on extensive measurements through that industry project in the harsh environment on the Norwegian Ekofisk field, it became apparent that the dynamic deflections of the platforms had to be incorporated in the design work. The industry standards of the day ignored this effect, and the initial reaction by the industry was somewhat reluctant to add the dynamic, as jack-ups were mostly used in shallow waters.*

Veritas Report no. 83-0145 explored dynamics of jackup platforms, so jackup designers learned how the design principles employed in 1974 could be improved on with the thinking assembled in 1985.

In the course of the FPSO boom in the North Sea in the 1990s, I had the good fortune to get to know Carl Arne Carlsen two decades after ETA. Both of us were speakers at FPSO conferences, but we provided different frames of reference; his from the DNV and mine from the contractor and designer. I did not know that he had been DNV's expert on jackup business back in the days of the ETA Europe Class. Just as he had been jackup authority, he later became an authority on FPSOs!

Years later DNV commented in 2014:

> *Building on these full scale measurements Carl Arne Carlsen and his colleagues developed analysis methods to implement these dynamic effects in the DNV MODU rules in 1982, documented in Veritas Classification*

Note 31.5 in 1984. With support from the Society of Naval Architects and Marine Engineers (SNAME) these aspects are today part of the industry standard for safe operations.

In 2014, OTC recognized DNV GL's Dr. Carl Arne Carlsen with a Distinguished Individual Achievement Award, recognizing his "outstanding, significant, and unique achievements, and extensive contributions to the offshore industry."

Dyvi Beta And *Dyvi Gamma* Work At The Historic Deck-Raising At Ekofisk (1987)

Dyvi Gamma was again employed at Ekofisk. Picture 24 is unusual for multiple reasons: (1) it shows a bright, clear, sunny day at Ekofisk, (2) both *Dyvi Beta* and *Dyvi Gamma* are there at the same time, and (3) it is taken during the time of raising all the decks at Ekofisk by six meters in August 1987.

It was a tremendous achievement to raise so many platform decks by that distance in that part of the North Sea, so the event was well-documented at OTC and elsewhere.

Dyvi Gamma had been converted to accommodation service by this time and can be seen higher up in Picture 24, on the right of the central walkway, located next to what is platform 2/4 T, i.e. the big concrete storage tank. Both units were needed to provide accommodation for the many workers, fabrication, and work space on deck. *Dyvi Beta* appears further down, on the left of the central walkway, next to platform 2/4 Q.

Picture 24:
Dyvi Beta and *Dyvi Gamma* at work at the Ekofisk complex in August 1987 in the Norwegian sector of the North Sea
Source: www.shipsnostalgia.com

Deliveries Of Jackups Of ETA Design After 1976-1977

The ETA Robray 300 class and ETA Europe Class jackups were controversial at the time. They challenged the original Big Four jackup designer-builders, stretched design methods, stretched overall jackup performance and load carrying capabilities, and used cast steel joints! When the second delivery wave for jackups started in 1978, ETA was out of business, so there was no one to promote these new designs and no engineering team to back up and work with builders. ETA had missed the boat on the 1978-1986 building boom in jackups!

From shipyard:	CFEM France	Hitachi Japan	Robin Singapore	ETA design?
Jackups delivered in 1976-1977: (ETA still in business)	2	2	2	All were ETA design
Jackups delivered in 1978-1986: (ETA no longer in business)	9	23	3	2 at Hitachi, 3 at Robin

Table L:
Jackup deliveries from the three builders of ETA jackups, during and after ETA

Still, ETA designs had a track record with three builders: the two experienced offshore industry fabricators (CFEM and Hitachi) and the new startup (Robin Shipyard), all as shown in Table L.

It is still perplexing why Robin Shipyard did not perform better, since they started on jackups relatively early in the industry (1973). They delivered their five ETA Robray 300 jackups, in 1976-1982 in contrast to Hitachi's 23.

Deliveries Of Smaller Jackups For Pool Company: Ten Built (1976-1982)

Pool Company retained ETA as jackup design consultant to develop designs to suit their particular well service, workover, and production drilling requirements. Pool chose the drilling-related equipment and systems, and ETA integrated these into the designs. Unlike the designs created, funded, and developed at ETA's risk, these designs were fully funded by ETA's client and belonged to Pool Company. They were for much smaller jackups than the 25,000-30,000 ft. maximum drilling depth exploration jackups that ETA had worked on in the previous years, but they were interesting in their own ways.

The Pool Company jackups were in two series: (a) the Pool 50 series, a workover unit for up to 90-ft. water depths in the Gulf of Mexico, and (b) the Pool 140 series, consisting of production drilling and workover units for up to 150-ft. water depths in the Middle East.

Pool 50 drew on what I'd learned in 1968, when I diagnosed the loss of the *Dresser II* workover jackup in South Pass 29 in the Gulf of Mexico. *Dresser II* also had had tubular legs and was mat-supported. The shear strength of the soil it had been set down on was very low, so it took little in the way of wind and wave when added to the dead weight and workover loads to flip that jackup. It sounds obvious today, but it was important to call out the bearing pressures imposed by the jackup's mat and know in advance whether soil properties would allow safe operation!

This industry lesson was helpful when applied to this new-mat supported jackup for comparable service, translating into operating book requirements to know well in advance the soil shear strength at each location before committing to work there.

Diagram 41 shows the general arrangement of the first of the Pool 50 series of jackups. The equipment arrangements were different and the design loadings were far lighter than those on the deepwater ETA Robray 300 Class and ETA Europe Class units. The Pool 50 series jackups used a cantilever configuration for getting over production platforms for well service and drilling. It had four tubular legs with diametrically opposed racks, elevated by a rack and pinion system. BLM and National Supply elevating systems were considered, but ultimately the first Pool 50 jackup was contracted for construction at Baker Marine in Ingleside, Texas and a hydraulically driven rack and pinion elevating system from Baker Marine Corporation (BMC) was used.

Diagram 41 shows the general arrangement of the Pool 50 series jackups. Picture 25 shows the first of the Pool 50 series units, delivered in 1976.

The Pool 140 series of jackups had a different set of drivers from the Pool 50 series; they were designed for about 150 ft. of water in the Middle East for production drilling and workover service and a maximum drilling depth of 12,000 ft. Diagram 42 shows an outboard profile and general arrangement.

Elevation

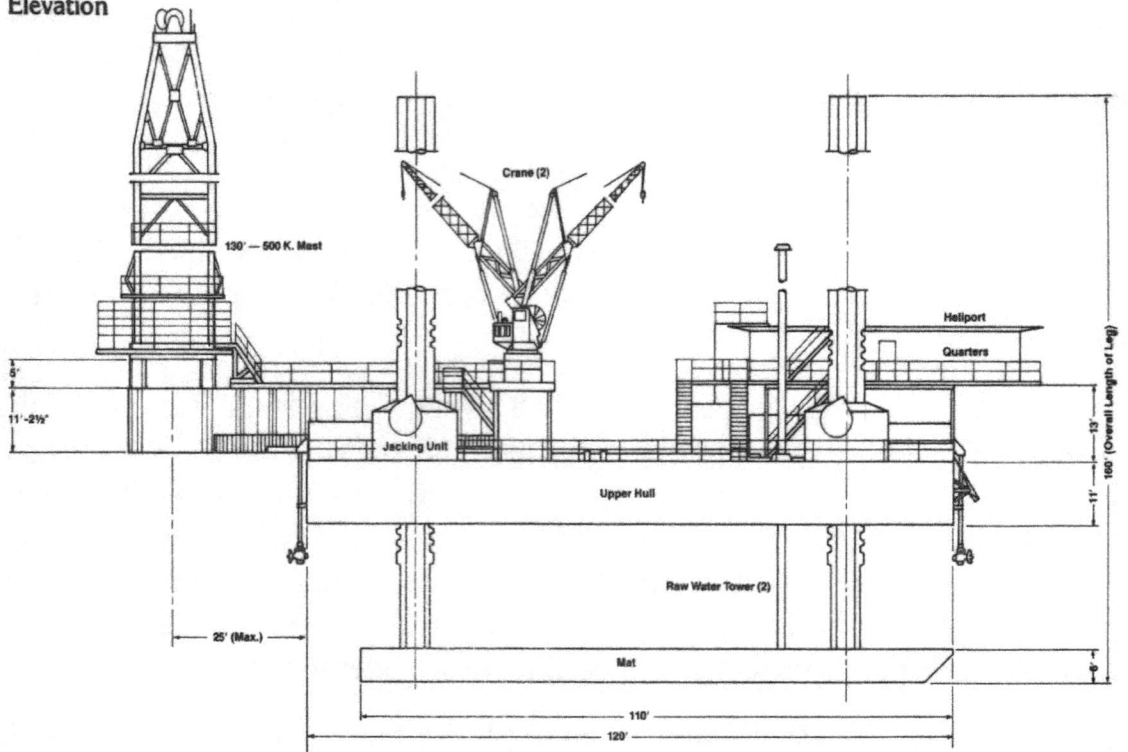

Plan

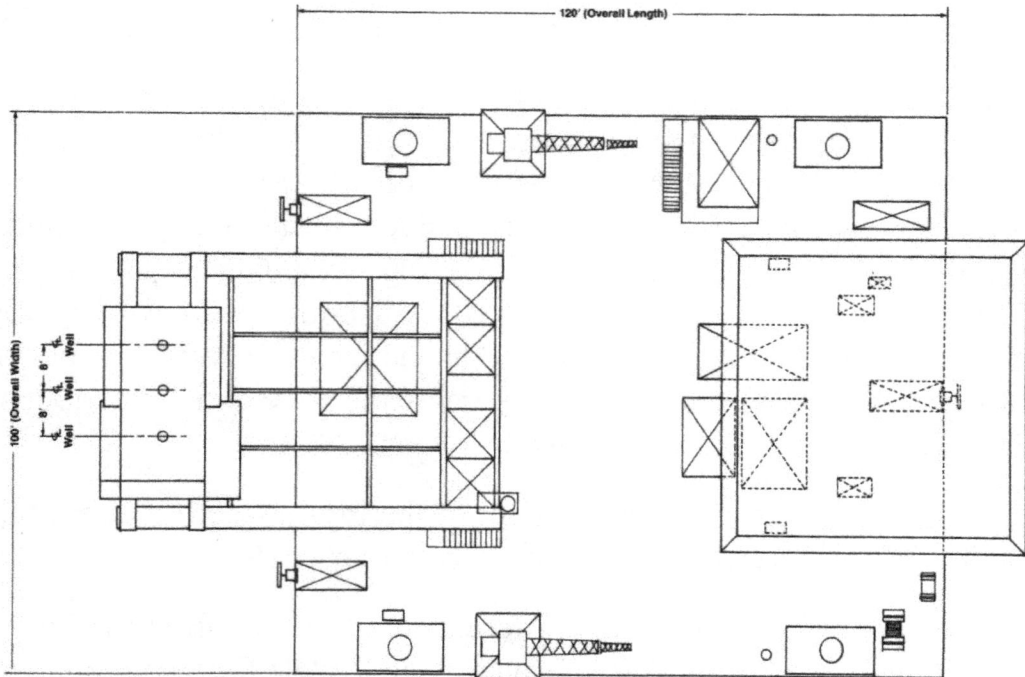

Diagram 41:
Typical arrangement of the Pool 50 series of jackups
Source: Pool 50 Christening Booklet, *1976*

Picture 25:
Pool 50 series of jackups: *Pool 50* in 1977
Source: Waiman Kwan

Unlike the Pool 50 series, this design employed three independent truss legs. But like the Pool 50 series design, it also employed a cantilever configuration for the same drilling and well-servicing business lines that Pool Company pursued.

In order to make the leg design as cost effective as possible and compete with existing jackups in similar service, ETA's engineers hit on the idea of having the rack on only one chord instead of all three. This was novel and reduced leg fabrication costs. It also reduced the investment in elevating equipment since jacks would only be installed at one chord of each leg instead of all three. There was a drawback: the leg design now had to accommodate a new distribution of loadings as only one chord could transfer vertical loads from the hull, making the structural design quite different from past designs.

The weights to be lifted were the same, but the controls and wiring for the electrically driven jacks were significantly less. This simplification provided savings in CAPEX.

Pool 142 was the first in the series and was designed from late 1974 into 1975. It employed jackhouses and National Supply rack and pinion jacks in a configuration generally like the ETA Robray 300 Class but smaller. The leg chord with opposed racks was the ETA design shown in the patent.

Pool 142 was built in Denmark at the Pars Dan yard. Subsequent jackups in the Pool 140 series were built in Singapore at FELS: *Pool Arabia 143, 144*, and *145*. Each had their own differences; the jacks came from Baker Marine in 143 and 144 and then from BLM for *Pool Arabia 145*. Jackhouses were not needed with these jacks since they were framed together with the leg guides and the hull. An idea of the the general arrangement is in Diagram 42.

Diagram 43 illustrates another difference: the change to the leg chord. Pool Company patented the leg arrangement in a filing in May 1977 after Pool 142 (the first of the 140 series of jackups) had already entered service and right in the middle of the heated legal dispute between ETA and Pool over the Pool 50 design (as elaborated in Part II).

The new leg chord with the rack was very simple: a T-section with the double-edged rack on the horizontal and the "vertical" of the pointing inward. Structurally it was not as efficient as the original ETA design, and wave and current loadings would be a bit more. But for this water depth and environment, it overall made sense.

Stress concentrations in that alternate leg design would be higher, hence the risk of fatigue cracking might have been greater, factors that may have emerged in the leg damage after the dry tow of *Pool Arabia 145* from shipyard to first location. Picture 26 shows *Pool Arabia 145* which is generally similar to *Pool Arabia 143* and *Pool Arabia 144*.

While the Pool jackups add to the count of jackups built that were of ETA design, their value and size was obviously much less than the earlier exploratory drilling rigs rated for 300-350 ft. water depths. A total of six Pool 50 series jackups were delivered to the Gulf of Mexico and four Pool 140 series jackups were delivered to the Middle East.

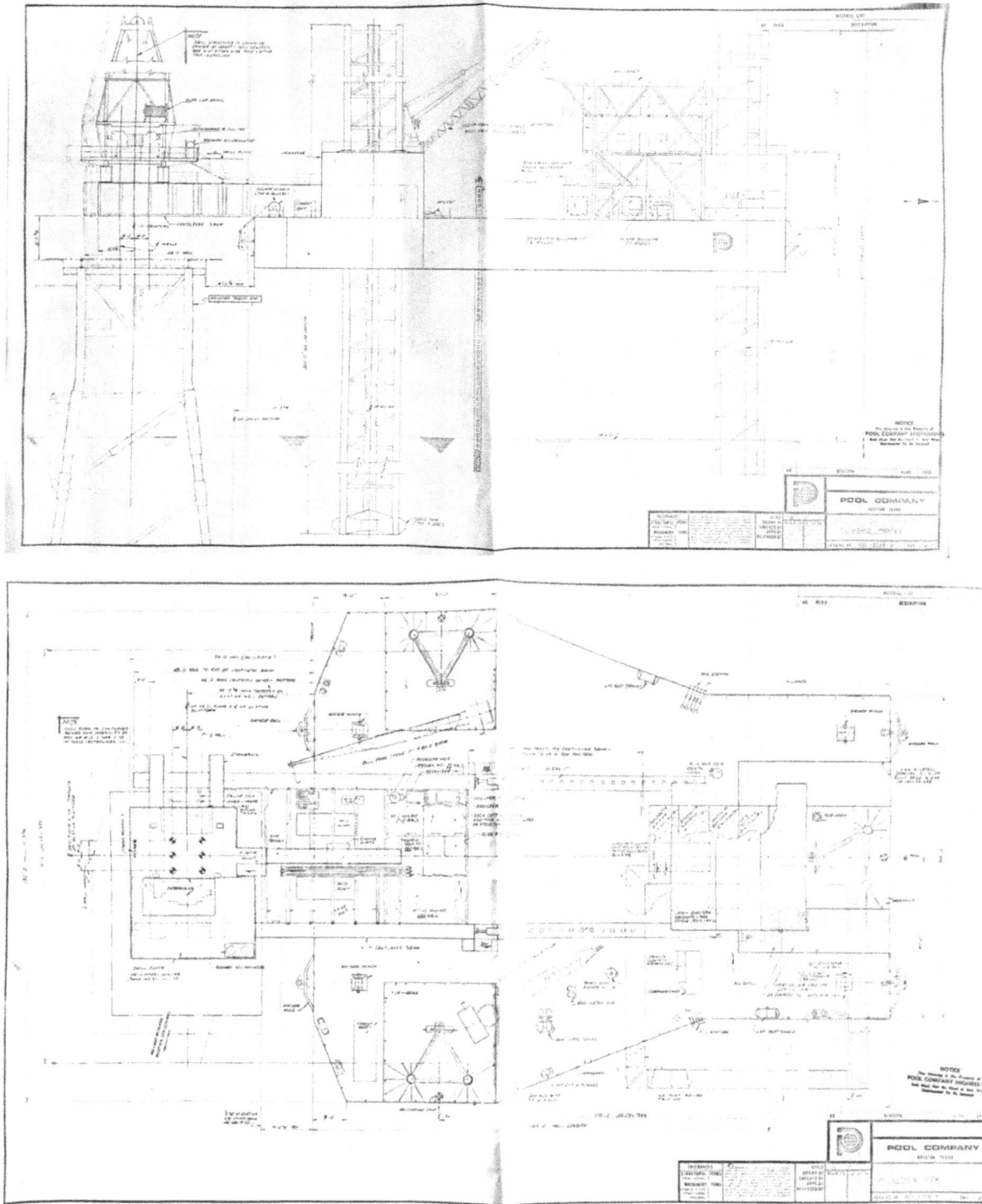

Diagram 42:
Outboard profile and main deck plan for the Pool 140 series of jackups
Source: Pool Company Drawings: "Outboard Profile" No. HO-5016-3 dated December 21, 1976
And: "Plan–Main Deck" No. HO-5016-1, dated December 21, 1976

Picture 26:
Pool 140 series of jackups: *Pool Arabia 145* in 1983
Source: Waiman Kwan

Nabors later acquired much of Pool Company's business and equipment, so –the names for the original 50 and 140 series of Pool jackups have changed in the listings in Table O below.

United States Patent [19]

Armstrong

[11] **4,160,538**

[45] **Jul. 10, 1979**

[54] **LEG STRUCTURE FOR JACK-UP PLATFORM WITH SINGLE POINT JACKING**

[75] Inventor: **James E. Armstrong**, Crosby, Tex.

[73] Assignee: **Pool Company**, Dallas, Tex.

[21] Appl. No.: **792,929**

[22] Filed: **May 2, 1977**

[51] Int. Cl.^{2} B66F 7/12
[52] U.S. Cl. .... 254/89 R; 254/95
[58] Field of Search .... 254/89 R, 95–97, 254/105; 61/91, 90; 405/196, 199

[56] **References Cited**

U.S. PATENT DOCUMENTS

3,332,663	7/1967	Cargile	254/107
3,606,251	9/1971	Willke et al.	254/95
3,743,247	7/1973	Willke et al.	254/95

Primary Examiner—Robert C. Watson
Attorney, Agent, or Firm—Richards, Harris & Medlock

[57] **ABSTRACT**

A trussed leg structure for use in jack-up mobile off-shore platforms of the type that include a plurality of support legs extending into the water down to the ocean floor with jacking units engaging each leg to raise and lower the platform structure relative to the surface of the water. The leg structure disclosed includes a plurality of mutually parallel and laterally spaced apart column members rigidly interconnected to define a triangularly-shaped leg. The leg structure has only a single elongated rack carried on one of the column members to provide single point jacking of the leg. The rack comprises a flat, plate-like rack member having a set of rack teeth extending along each edge for meshed engagement with a separate one of a pair of pinions on a jacking unit. Two of the three column members are made of tubular pipe. The third column member, which carries the elongated rack, is an elongate rack plate oriented perpendicular to a plane defined by the first and second column members. The rack plate in combination with the rack member gives the third column member a generally T-shaped configuration. Cross-bracing extends between adjacent column members and includes horizontal brace members with diagonal brace members extending between adjacent pairs of horizontal brace members.

11 Claims, 3 Drawing Figures

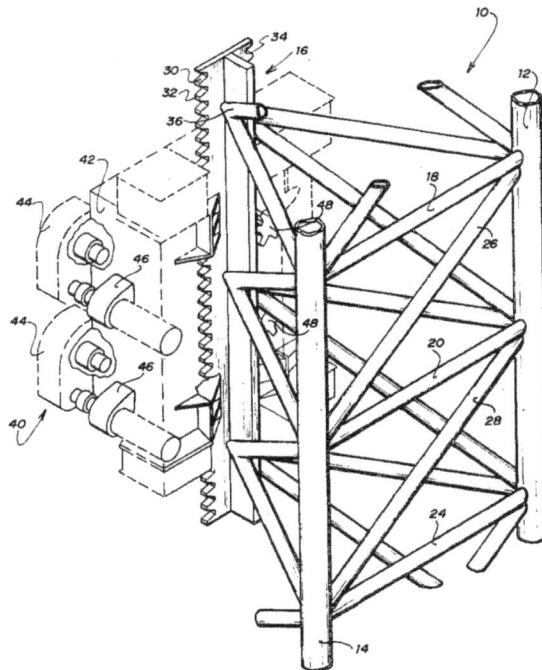

Diagram 43:
Pool leg rack chord design on Pool 143, 144, and 145 jackups

Particulars On All Jackups Built Worldwide From Designs by ETA

ETA's career total came to 2+9+1+4+6=22 jackups, as summarized in Table M, where they are ranked by water depth. Table M also shows the current status of these units as best can be determined from checking multiple sources. At the end of 2017, the breakdown in Table M shows 7 working, 3 stacked, 4 converted to other services, 4 lost, 3 scrapped and 1 unknown.

The next two tables pull together information on all the jackups built from ETA's designs with more detail: jackup names, when they were built, and their current status as far as can be determined from at least three databases online and information from industry colleagues.

The compilation of data started with *Rigzone*, but was difficult because of the passage of time and locations when jackups often were in China or somewhere else where data was difficult to verify. Cross-checks were made against Subsea, Infield, and Braemar, and I attempted further online searches to confirm data on specific jackups. Sometimes no trace could be found for what happened to the jackups in the Pool fleet; those jackups are labeled "unknown." Sometimes the lack of data for jackups built in the late 1970s and early 1980s corresponds to them being lost in accidents— or more likely retirement in the 1990s, in which case no record of their demise can be found in the databases.

We were fortunate to have access to the World Offshore Accident Database (WOAD), which DNV-GL Ltd. has meticulously built up over the years. It gave yet another cross-check on jackup fleet data, as well as histories of incidents, from very minor spills to complete constructive total loss data.

The jackups listed in Table N are all of designs in which ETA took the initiative and risk in their development when we came up with the design concepts. Table N thus shows how, of the nine jackups built using the ETA Robray 300 Class design, one was lost in 1985 and the eight remaining are still in existence in 2017.

Max. water depth, ft.	Totals built	Design		Reported status in 2017
350	2	ETA Europe Class	1	Conversion (MOPU), working
			1	Lost in tow accident in 1990
300	9	ETA Robray 300 Class	5	Working
			1	Conversion to accomodation unit
			2	Stacked
			1	Lost in blowout in 1985
200	1	ETA Europe Class variant	1	Stacked in UK waters
150	4	Pool 140 series	2	Working
			1	Converted accomodation unit 2014
			1	Scrapped
90	6	Pool 50 series	1	Conversion to liftboat 2015
			2	Lost: blowout, capize 1982, 1987
			2	Sold for scrap, 2011, 2015
			1	Unknown
	22			

Table M:

Summary of jackups built using ETA designs and status at the end of 2017

No.	Jackup Name	Latest Owner	Builder, Location	Operator / Region	Year delivered	Year retired	Max. water depth, ft.	Jacking System	Status at end 2017
								(rack and pinion) electric or hydraulic drive	

First Group: Designed in 1973, the "ETA Robray 300 class" design was for Far East service, with three independent truss legs, triangular, design length 425 ft., with cast steel joints, ABS classed, max. water depth 300 ft., 25,000 ft. drilling depth

No.	Jackup Name	Latest Owner	Builder, Location	Operator / Region	Year delivered	Year retired	Max. water depth, ft.	Jacking System	Status at end 2017
	Built by Hitachi in Japan for Robray Offshore Drilling and China Offsoue Services Ltd. (COSL)								
1	*Yu Son* (ex *Robray JU5*, ex *Ednastar*, ex *COSL 935*)	Korea Oil Exploration Co.	Hitachi, Japan	- - - China	1976	- - -	300	National Supply elec	41 years old, stacked in China
2	*Bohai IV* (ex *Robray JU4*, ex *Ednarina*)	CNOOC	Hitachi, Japan	CNOOC, China	1977	- - -	300	National Supply elec	40 years old, drilling, Bohai Bay
3	*Nanhai III* (ex *South Seas III*)	COSL (ex PRC Marine)	Hitachi, Japan	P R C	1980	1985	300	National Supply elec	Blowout - Malaysia
4	*Nan Hai IV*	COSL	Hitachi, Japan	CNOOC - China?	1980	- - -	300	National Supply elec	37 years old, drilling, Bohai Bay
	Built by Robin Shipyard in Singapore for COSL, Odebrecht and ONGC *without* cast steel joints)								
5	*Nan Hai I* (ex *South Seas I*)	COSL (ex PRC Marine)	Robin Shipyard, Singapore	CNOOC China	1976	- - -	300	National Supply elec	41 years old, drilling
6	*Kan Tan II*	COSL (ex PRC Marine)	Robin Shipyard, Singapore	- - - China	1977	- - -	300	National Supply elec	40 years old, ready stacked
7	*COSL Rigmar* (ex *Norbe 1*, ex *Port Rigmar*, ex *RigMar 301*)	COSL (ex PRC Marine)	Robin Shipyard, Singapore	- - - North Sea	1979	- - -	300	National Supply elec	Converted 1992 to accommodation unit (Bethlehem/Breivik), stacked in Dutch waters
	Variants on ETA Robray 300 class - cantilever conversion, 1 shorter legs, 1 longer legs								
8	*Sagar Gaurav*	ONGC	Robin Shipyard, Singapore	ONGC, India	1982	- - -	350	ETA BLM elec	35 years old, drilling/workover
9	*Sagar Shakti*	ONGC	Robin Shipyard, Singapore	ONGC, India	1982	- - -	350	ETA BLM elec	The newest: now 35 years old, drilling

Second Group: Designed in 1974, the ETA Europe Class design was for North Sea service, with three independent truss legs, triangular, design length 508 ft., using cast steel joints, DNV classed, max. water depth 350 ft., 30,000 ft. drilling depth. When delivered they were the biggest and harshest environment jackups in the world and represented ETA's greatest jackup design accomplishment

No.	Jackup Name	Latest Owner	Builder, Location	Operator / Region	Year delivered	Year retired	Max. water depth, ft.	Jacking System	Status at end 2017
10	*Baltic Beta* (ex *West Beta*, ex *Dyvi Beta*)	Petrobaltic	CFEM, France	Petrobaltic North Sea	1976	- - -	350	ETA BLM elec	41 yrs old, 2015 as MOPU in Poland
11	*West Gamma* (ex *Dyvi Gamma*)	Dyvi Drilling A/S	CFEM, France	North Sea	1977	1990	350	ETA BLM elec	Lost under tow in North Sea 21Aug90
	Variant on ETA Europe Class - cantilever conversion, shorter legs								
12	*Paragon 391* (ex *Noble Julie Roberston*, ex *Ocean Scotian*, ex *Arethusa Scotian*, ex *Zapata Scotian*)	Paragon Offshore	Promet, Singapore	- - - North Sea	1981	- - -	200	Baker Marine hyd	Ready stacked at Harwich, UK

Table N:
List of jackups built worldwide from designs by ETA, in which concept originated with ETA

For the two jackups of the ETA Europe Class design, one was lost in 1990, and the other is still working in the North Sea today, now as a MOPU. A variant built originally for Eastern Canada service is still in existence and is stacked.

In contrast to Table N, Table O describes jackups designed to fit a concept provided by an ETA client, i.e. Pool Company. Of the ten jackups designed for Pool Company, Table O shows how only four are known with reasonable certainty to still be in existence today.

No.	Jackup Name	Latest Owner	Builder, Location	Operator / Region	Year delivered	Year retired	Max. water depth, ft.	Jacking System (rack and pinion)	Status at end 2017

Third Group: Designed in 1975-1976, much smaller jackups where ETA performed basic jackup design work commissioned by Pool Company on their needs for two sizes: (a) for development drilling (Pool 140 series), 3 independent truss legs, 150 ft. max. water depth, 12,000 ft. drilling depth), and (b) for workover service (Pool 50 series), 4 tubular legs with a mat, max. 90 ft. water depth. No cast steel joints in any of them.

No.	Jackup Name	Latest Owner	Builder, Location	Operator / Region	Year delivered	Year retired	Max. water depth, ft.	Jacking System	Status at end 2017
13	*Pool Rig 50*	Nabors	Baker Marine, USA	US GoM	1976	2011	90	Baker Marine hyd	Sold for scrap
14	*Pool Rig 51*	Pool Company	Baker Marine, USA	US GoM	1979	? ? ?	90	Baker Marine hyd	Unknown
15	*Pool Rig 52*	Pool Company	BMC, USA / FELS, Singapore	US GoM	1979	1982	90	Baker Marine hyd	Blowout, fire. GoM
16	*Pool Rig 53*	Pool Company	BMC, USA / Amardah, S.Africa	US GoM	1983	2015	90	Baker Marine hyd	Sold for scrap
17	*Pool Rig 54*	Pool Company	BMC, USA / Amardah, S.Africa	US GoM	1982	- - -	90	Baker Marine hyd	Sold 2015 as liftboat for Nigeria
18	*Pool Rig 55*	Pool Company	BMC, USA / Amardah, S.Africa	US GoM	1982	1987	90	Baker Marine hyd	Capsized, scrapped
19	*Pool Arabia Rig 142*	Pool Company	Pars Dan, Denmark	Middle East	1976	1992	150	National Supply elec	Dismantled, retired 3/92
20	*Nabors 655 (ex Odin Star, ex Pool Arabia*	Nabors	FELS, Singapore	Middle East	1979	- - -	150	Baker Marine hyd	Drilling
21	*Foresight Driller V (ex Odin Moon,*	Foresight	FELS, Singapore	Middle East	1980	- - -	150	Baker Marine hyd	Sold 2014 as accommodation unit
22	*Realm I (ex Pool Arabia Rig 145)*	Foresight	FELS, Singapore	Middle East	1982	- - -	150	ETA BLM elec	Drilling

<u>Table O</u>:
List of jackups built worldwide from designs by ETA, in which concept originated with client

PART VI

ALL-TIME JACKUP BUILDING BOOM, SPREAD OF JACKUP DESIGN SOURCES WORLDWIDE, LONG STAGNATION, AND UNPRECEDENTED OVERSUPPLY

Throughout my research of what happened to the jackups designed by ETA, it was obvious that the pattern of orders for new jackups has fluctuated a lot year to year, and the ETA years were no different. Looking back, the pattern of deliveries was significant in how the jackup business changed during 1970-1986 and how the number of design sources grew and grew, as did the shipyards that built the jackups.

Tracking design sources was important because it led to uncovering a pattern not seen in the tracking of owners or builders. A *design source* might be a builder, an owner, or an independent design firm such as ETA or Friede & Goldman. For purposes of analysis, delivery was taken for timing rather than order date.

Data from *Rigzone* on jackup deliveries over 1970-1986 were examined, counting by design source to discern patterns of jackup orders from different design sources in this market.

Worldwide jackup deliveries during 1970-1986 were then plotted in Diagram 44 for the two ordering cycles. These ordering cycles resulted in the smaller 1970-1977 wave of jackup deliveries, which coincided with the years ETA was in existence and then the second, the all-time jackup-building boom cycle of 1978-1986, after ETA had disappeared.

Bethlehem, Levingston, Marathon LeTourneau, and The Offshore Company clearly dominated the jackup design and construction business back in 1970 when ETA started. They continued to dominate as design sources throughout the 1970s, but they started to have a decreasing market share after about 1977.

As demand for jackups grew worldwide, they would continue to build in their own yards in the U.S. and then build in yards in Singapore (Bethlehem and Marathon LeTourneau).

Levingston made a management agreement in Singapore during 1969-1972 with Far East Shipbuilding that changed its name to become Far East Levingston Shipbuilding (FELS) but Levingston did not build any jackups of their design there, choosing instead to license Mitsui in Japan. FELS reportedly lost money under Levingston but subsequently forged a truly remarkable record on their own in becoming one of the world's leading MODU builders. The drilling contractor member of the original Big Four (The Offshore Company) would design its jackups and select a yard to build them, sometimes additionally licensing other shipyards to build their designs for other owners.

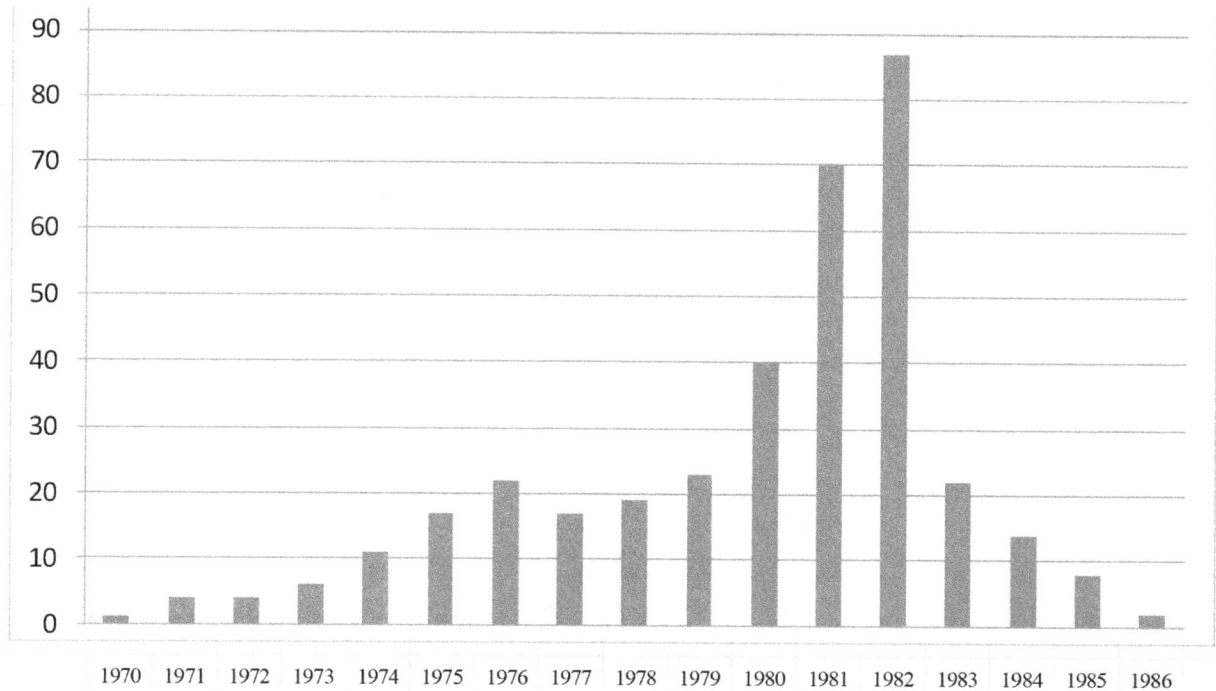

Diagram 44:
Plot of annual jackup deliveries during 1970-1986
Source: Derived from Rigzone *data*

Diagram 44 shows how jackup deliveries grew dramatically in the second wave to be followed by an abrupt downturn. But that is only part of the story.

The underlying data shows that the ordering pattern and activity by jackup <u>design sources</u> fell into two waves. In general, deliveries were roughly two years after orders. The ordering surge of 1973-1974 was followed by a dip in orders until a recovery took hold about 1978. There was particularly intense ordering during 1978-1980.

Industry Upheavals (1970s and 1980s)

All these jackups being built were happening in an era of great change in the offshore drilling world and more broadly in the petroleum industry itself. Living it, life was uncertain, often exciting. In recent years industry leaders have written about the patterns of the upheavals.

Robert F. ("Bob") Bauer (1918-2011), was founder and CEO of Global Marine, a real pioneer. He started the company in 1959, a leader in developing the CUSS series of drillships and the Deep Sea Drilling Project with the *Glomar Challenger*. He guided the CIA in how they recovered the K 129 Russian submarine with *Glomar Hughes Explorer* in 1972. He retired in 1981 and in his 2008 autobiography he wrote:

> *"The decade of the 1970s was one of the most unsettled periods in the petroleum industry's history and its span of years had an indelible effect on the offshore drilling and production segment. Global Marine's fortunes vaulted from boom to bust and back to boom in only a few years. Ultimately this led to the undoing of the company I had founded decades earlier and had always envisioned as one that would continue to prosper as an independent enterprise."*

Global Marine went into Chapter 11 bankruptcy in 1986 (and later emerged successfully) with its shares plunging from an all time high of $73.00 to $0.15. It must have been an terrible experience for Bob Bauer, the company's founder, to see it get to this after the decades of building. He had retired five years earlier and a new management strategy had been installed for growth but was in the wrong direction.

The 1980s were described concisely and dramatically in the 2011 book: *Deepwater Petroleum, Exploration and Production* by three Shell movers and shakers (William Leffler, Richard Pattarozzi and Gordon Sterling):

> *"In the mid 1980s, ominous dark clouds floated over the Gulf of Mexico. OPEC learned that with $34 they had prices themselves out of many markets. Consumers found unprecedented ways not to use oil in their cars, industrial plants and buildings. The oil price softened to $28 in 1983 and collapsed to $10 in 1986."*

ETA was a tiny outfit in comparison, but it had its parallels with much bigger companies and their leaders that had pioneered groundbreaking developments and then had faced the booms and busts in the offshore drilling business.

Design Sources Spread Worldwide

The original Big Four did well in the second wave: 136 deliveries versus 67 in the first wave. They accomplished this success by building overseas in addition to the U.S. Worldwide jackup demand in the second wave was far higher than in the first wave. It stimulated five new builder–designers worldwide, plus five new independent design firms. Table P shows the numbers of jackups delivered by each design source.

Four of the five new independent design firms were U.S.-based, and one (MSC) was in Europe. MSC subsequently became part of Gusto, which had been designing jackups even before ETA came along. While the original Big Four clearly dominated deliveries during the first wave 4:1, they lost that dominance in the second wave (0.9:1), even though they about doubled their total deliveries. The world demand for jackups was simply that huge.

As Far East yards became skilled in the design and construction of all kinds of MODUs and not just jackups, yard efficiencies grew and their building costs were less than in the U.S. Then the deepening petroleum industry downturn affected everyone, making life extra difficult for the traditional MODU builders around the U.S. Gulf Coast.

By the low point of 1986, these historic economic trends had led to Levingston Shipbuilding going out of business, and Bethlehem Steel in Beaumont followed soon after in 1989. The Offshore Company struggled through the downturn and engaged in more deepwater drilling work, as opposed to jackup business. LeTourneau was the only one of the original Big Four that continued but in a lower scale. However, it eventually closed its Vicksburg Mississippi yard.

The deliveries by design source in Table P show how the two waves of 1970-1977 and 1978-1986 reflect the shift from the small U.S. club of design sources (the established Big Four) to essentially a worldwide sourcing of jackup designs and construction.

The industry trend towards increased use of rack and pinion elevating systems and away from hydraulic jack and pin systems were underway in the first wave and continued; about 75% of deliveries in the first wave were rack and pinion, increasing to 83% in the second wave, and close to 100% today.

Original Big Four jackup designers		Other shipyard-designers worldwide		Independent design firms		Totals, all design sources
A		B		C		A+B+C

- - - - - The first wave of 1970-1977 - - - -

16	Bethlehem	1	Baker Marine	1	CDB Korall	
9	Levingston	3	IHC Gusto	12	ETA	
37	Marathon Letourneau					
5	The Offshore Company					
67		4		13		**84**

Domination of the Original Big Four was 4.5:1   (A / (B + C)

- - - - - The second wave of 1978-1986 - - - -

42	Bethlehem	44	Baker Marine	6	CDB Korall	
11	Levingston	9	CFEM	1	Donhaiser Marine	
76	Marathon LeTourneau	3	Dalian Shipyard	10	E T A	
7	The Offshore Company	22	Hitachi	34	Friede & Goldman	
		6	IHC Gusto	3	M S C	
		2	Mitsubishi	2	PanX	
		9	Mitsui	2	Penn Engineering	
136		95		58		**289**

Domination of the Original Big Four was no more: 0.9:1   (A / (B + C)

- - - COMBINED: Total deliveries of new jackups from all design sources during 1970-1986 - - -

203	**99**	**71**	**373**

<u>**Table P**</u>:
Changing patterns of jackup deliveries from different design sources
during 1970-1986

The two big winners in the second wave were Baker Marine in the "Other shipyard-designers worldwide" category, with 44 deliveries, and Friede & Goldman in the

"Independent design firm" category, with 34 deliveries.

In the second wave, new names joined the ranks of the jackup design sources:

Shipyard-designers: CFEM, Dalian, Hitachi, Mitsubishi, Mitsui

(CFEM, Hitachi and Mitsui had already built jackups designed by others in the previous wave)

and:

Independent firms: Donhaiser Marine, Friede & Goldman, MSC, PanX, Penn Engineering.
(All were newcomers to the jackup design business)

The huge growth in the world's jackup fleet that occurred in the second wave has not been repeated since. 1986 saw the start of a radical slowing in jackup deliveries that has continued to today. Although there have been spurts in deliveries, they have been nothing like Diagram 44's pattern of deliveries.

As the original Big Four largely disappeared after 1986, in the subsequent three decades of 1987-2017, new jackup design sources such as KFELS and Zentech emerged.

Major Changes In The World's Jackup Fleet Growth

Table Q is a revealing comparison of the world's jackup fleet at three snapshots in history: 1974, 1985 and 2017.

The data comes from different sources, has not been rigorously checked for consistency (that would be practically difficult), but does feel right. It uncovers surprising historical growth patterns of the world's jackup fleet:-

(i)   In the 11 years of 1974-1985, even recognizing that the world's jackup fleet has always grown in spurts, the average net growth was about 21.7% p.a.

(ii)   There were no dramatic changes during the next 32 years of 1985-2017, when the jackup fleet was generally stagnant at an average net growth of about 1.4% p.a. However, in the short term, it was volatile year to year.

Year:	1974	1985	2017
Market condition:	Some stability, sanity (?)	First really big downturn	Biggest ever downturn
Jackups working:	120	331	292
Jackups without contracts:	10 (estimate)	109	349
Total jackup fleet:	130	440	641
Jackup utilization:	92.3%	75.2%	45.6%
Idle jackups, delivered or still under construction:	none	none/few	107
Source:	ETA, Offshore 1974	Offshore Engineer 9/1985	Bassoe Offshore 11/2016
Growth of jackups at work:	- - -	+211 in 16 years	-39 in 32 years
Overall jackup fleet growth:	- - -	310 in 11 years (av. 21.7% p.a.)	201 in 32 years (av. 1.4% p.a.)
General trends:	Major true fleet growth in first 11 years, capability improvements over next 32 years, but flat fleet growth		
ETA history:	ETA Europe class design contract signed	ETA had gone out of business in 1977	ETA Saga published
Author history:	President, ETA	"Staying alive in '85"	Asymptotic retirement!

Table Q:

Three key snapshots of jackup fleet growth from ETA's time to today

(iii)   The speculative binge of the last few years resulted in an oversupply in 2017 like never seen before.

Throughout all these years, the normal process of fleet attrition continued via rig retirements and occasional losses. Attrition was small in 1974 with a relatively small and young fleet. In the last few years, attrition has grown and has varied widely from year to year as Table R indicates. However, fleet reductions were never enough to forestall the oversupply and crashing day rates.

Long term, the utilization of the world's jackup fleet deteriorated from about 92% in 1974 to something like 75% in the dip of 1985 to only 46% in 2017, a disquieting historical trend.

It is remarkable that there are fewer jackups working in the 2017 downturn than there were working in the last major downturn of 1985. There is a record surplus of jackups in 2017.

The challenge now becomes finding something to do with all these jackups idle for many years. It's quite opposite from the time when ETA saw a market need for new equipment!

Even the higher attrition levels of the last year or two (Table R) may not do much in reaching a balance with demand in the next few years, unless some unexpected petroleum industry upturn and a change in jackup retirement rates occur!

Going into 2017, there were 296 jackups in excess of the market balance figure of around 345. The world's jackup fleet faces tough times with an overcapacity of about 85%, based on Bassoe's figures from the end of 2016 with a start of strengthening starting now in 2018.

Year:	2000	2001	2002	2003	2004	2005	2006	2007	2008	2009	2010	2011	2012	2013	2014	2015	2016	Total
Attrition:	1	0	5	9	5	7	3	4	4	6	1	12	21	7	10	17	22	134

Data source: *Rigzone*   Average: 8

Table R:
Attrition in the world's jackup fleet during 2000-2016

In contrast, the FPSO business is unhappy with the number of idled FPSOs in its fleet. According to Fearnleys, another Norwegian offshore advisory, in 2017 there were about 22 idled FPSOs out of 218 (around 10%). In comparison with the idle jackup rates, the FPSO world has had it easy so far in this downturn!

A Barometer On Offshore Business

2018 saw the fiftieth Offshore Technology Conference (OTC) in Houston. It has become a remarkable gathering of offshore experts, business people, and individuals recognized as industry pioneers. It has maintained a high standard in all it does and stringent quality management year after year. Going to OTC has become something of a ritual in the worldwide offshore industry.

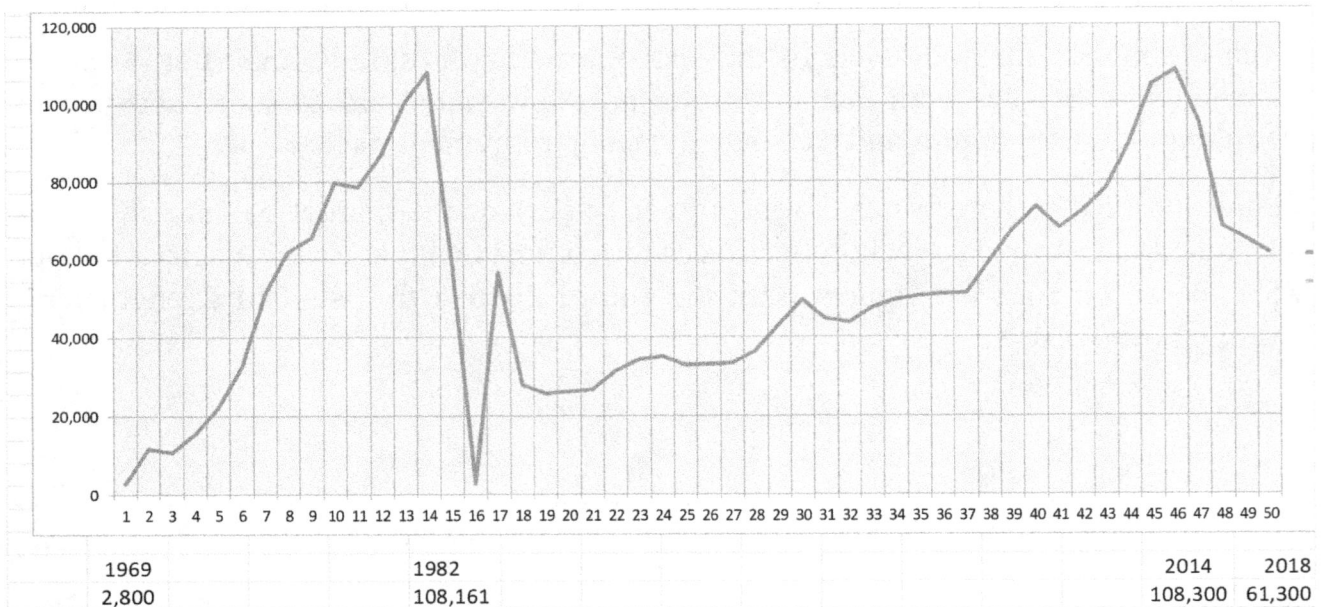

	1969	1982	2014	2018
	2,800	108,161	108,300	61,300

Diagram 45:
OTC attendance as a barometer for offshore business
Source: SPE, May 2018

This annual event has mirrored the offshore industry climate, reflecting the business atmosphere that ETA lived in the first half of the 1970s. Diagram 45 shows attendance at OTC since it started in 1969. The steep upward trend from 1970 demonstrates the booming interest in everything offshore, just when the world's jackup fleet was growing like never before with ETA at the front end of it.

That upward slope keeps on going until 1982's record of 108,161 attendees: growth corresponding to the years when jackups of ETA design continued to be delivered. Then the slide started.

We all saw the bad dip from 1982 into 1986. There was slow recovery as attendance erratically built up once more. In 2014, attendance was 108,300 attendees, back at the record level seen 32 years earlier. It was again short-lived; the bust started just weeks after that high point, and attendance four years later was about half at 61,200 with the 2018 attendees talking about a bottoming out.

In 1969 at what was then the Albert Thomas Convention Center in downtown Houston, the first OTC was held. For many years its venue has been the huge NRG complex. To celebrate the fiftieth anniversary, the organizers of OTC held a special luncheon for the few industry veterans that had come to OTC every year since it started. The organizers contacted individuals who they believed had been there all fifty years.

It was a surprisingly small number. There were eight including Peter Lovie, although admittedly I had skipped OTC in 2017 and 2018. Nevertheless the OTC organizers very graciously sent a lucite-encased medal to celebrate being one of their legacy members.

PART VII

CLOSING THOUGHTS:

LEARNINGS AND CONTRIBUTIONS

WORLD'S LARGEST JACKUPS THEN AND NOW

REFLECTIONS ON THE ETA ADVENTURE

Opportunities Taken And Missed, Lessons Learned

The jackup fleet growth shows how ETA had a truly unique opportunity. It detected the market need a year or two in advance and successfully acted on it, working on jackup designs for two new offshore drilling ventures seeking a competitive edge: Robray Offshore Drilling Co. and Dyvi Drilling A/S.

The extraordinary demand for jackups in the 1970s created the circumstances for ETA to enter this design field. At first, ETA was virtually the only independent design firm. ETA demonstrated how (i) the independent jackup designer business model was doable, and (ii) that their new generation of designs was up to challenging the original Big Four jackup suppliers.

ETA's timing was fortuitous; in 1970-1975, the world demand for jackups was starting to take off, and there was room for technical improvements. ETA seized the opportunity. In today's world such an opportunity would be virtually impossible.

The ambitious in the new generation in 2018 will have to find some business line other than jackup designs to get that lucky in today's world! Technicalities aside, a more important and truly American phenomenon was at play. Despite the vaunted conservativism in the petroleum industry and the boom and bust cycles, in the 1970s the climate of Texas' offshore drilling world made it possible for enterprising people to work hard and build a business. This encouraged people from other parts of the world to come to America to get accepted and become part of that business world. The American Dream does exist— then and now too.

The established U.S. offshore drilling contractors had done their share of pioneering and preferred not to go first! ETA's big jackup design clients were instead new overseas offshore drilling contractors seeking to gain an advantage over their U.S. counterparts via ETA's new jackup designs.

Going first at a shipyard with a new design is not a preferred modus operandi, since all the fabricating bugs may not have been taken out. To avoid problems, more planning and more engineering are often necessary, and still the risks of delay and overrun loom large. ETA saw how it takes time for operators, drilling contractors, and builders to come to accept a new generation of designs. In that regard, it was unfortunate that ETA was not around long enough to worry through the building and design trade-offs and have a chance of capitalizing on that learning process.

Initial Negative Shipyard Responses To New Designs

It was obvious from reactions that both Robin Shipyard and CFEM were not enthused to build the ETA designs. It did not seem to faze Hitachi. Obviously it meant figuring out how to build a design they had not built before— and having to use this new idea of cast steel joints in the legs on top of it all! We were inexperienced in ETA and likely needed to better explain the design novelties and supporting their efforts, despite our budget constraints.

In the long term, a truer measure of success was the longevity of these ETA Robray 300 Class and ETA Europe Class jackups, which remained in operation through the ups and downs in the offshore drilling business with many of them operational today. Another measure that demonstrates the effective design was the adoption of the leg chord design by other builders and used on at least 28 more jackups that were delivered in the several years post ETA (1977-1983).

The idea of cast steel joints was not so successful, despite their various advantages in structural design. The concept was realized as worthwhile by three major Japanese builders (Hitachi, Kawasaki and NKK), which offered to provide cast steel nodes. However, their marketing campaigns did not pay off; neither designers nor jackup builders of the day went for the idea.

ETA's pioneering encouraged other designers to enter the fray in the second wave of jackup deliveries during 1978-1986, in which worldwide deliveries of jackups were more than three times what they had been in the first wave during 1970-1977 (289 v. 84). In 1977, ETA's failure to survive meant losing any chance of benefitting from the momentum it had achieved in jackup designs by taking a share of the second wave.

The increasing age of the fleet of jackups of ETA design signals that they have performed well as designed. Still, they may soon be headed for retirement, just like their team of ETA designers.

ETA's Industry Contributions

Briefly, the significant and beneficial ETA accomplishments that stand out are suggested as:

1. Showed how a young independent design firm could succeed in the conservative jackup design business in competition with the Big Four established designer-builders of the day.

2. Created a new jackup design for 300-ft. water depths for Far East service (ETA Robray 300 Class), offering reduced steel weights, double the typical variables capacity, and reduced need to remove leg sections for ocean tows. Nine were delivered during 1976-1982.

3. Introduced the use of cast steel joints in jackup leg design to reduce joint stresses and enhance fatigue performance. Cast steel joints were used in six jackups delivered from two builders during 1976-1982.

4. Introduced an opposed-two-rack-on-center leg chord design that was structurally efficient, able to be widely fabricated, and used in at least thirty-nine jackups delivered during 1976-1986 from five builders.

5. The ETA Europe Class design, classed by DNV, pioneered harsh environment North Sea jackups. *Dyvi Beta* and *Dyvi Gamma*, delivered in 1976-1977, were the largest jackups in the world at that time. They prompted an industry movement that continued for more than a decade to learn more about the fundamentals of harsh environment jackups.

The World's Biggest Jackups Then And Now

Table S compares the design and capabilities of *Dyvi Beta*, which is still working, with the biggest jackup working today. The leader in North Sea jackup capabilities today is the *Maersk Intrepid* and its three sister jackups.

	THEN	NOW
General:		
Year delivered	1977	2015
Jackup name	*Dyvi Beta*	*Maersk Intrepid*
Design	ETA Europe Class	Gusto MSC CJ70-X150MD
Builder	CFEM	Keppel FELS,
Builder location	France	Singapore
Owner	Dyvi Drilling A/S	Maersk Drilling
Status in 2017	MOPU, N. Sea	Drilling, N. Sea
Hull and legs:		
Max. water depth, ft.	350	492
Leg length, ft.	508	678
Jacking system	rack & pinion	rack & pinion
Jacking speed, ft./min.	1.5	1.6
Hull dimensions, ft.	230 x 212 x 27	291 x 336 x 39.4
Classification	D N V	D N V
Accommodation	83 people	180 people
Variables capacity, kips	9,500	26,600
Drilling systems:		
Hook load, kips	1,300	2,000
Hull configuration	Traditional slot	"XY" Cantilever
Drilling depths, ft.	30,000	40,000
Drawworks, h.p.	3,000	5,750
Mud pumps	2 @ 1,600 hp x 5,000 psi	4 @ 2,200 hp x 7,500 psi

Table S:
The world's biggest jackups, then and forty years later

Picture 27:
World's biggest jackup in 2017: *Maersk Intrepid* afloat and ready to go on
location (left) and in transit on a dry tow (right)
Source: Maersk Drilling

ETA's engineers were naturally proud to design the world's biggest jackups for harsh North Sea conditions and meet the demands of stringent regulatory standards. Over forty years the world's biggest jackups have grown, with leg length increased from 508 to 679 ft., total variables capacity almost tripled (from 9,500 to 26,600 kips), living quarters more than doubled (83 to 180 people), and drilling depths advancing from 30,000 to 40,000 ft.

The advances in Table S speak for themselves but the table does not cover all the efficiencies in drilling operations now possible and the improvements in equipment and automated systems. Rig crew living condition and safety have advanced remarkably.

The huge leg length is very apparent in Picture 27 of the *Maersk Intrepid.* The image on the right dramatically shows the improvement in mobilization speed gained from the use of dry tows, which are often three or four times the speed of the wet tows of forty years ago.

Reflections Today On The ETA Adventure

With passing years, one comes to recognize how ideas, people, and organizations shape our lives and those of others. Fundamentally, family and upbringing shape so much. I was fortunate—growing up in a family in Scotland, encouraged to be the best you can possibly be and do what you really enjoy for a profession. It was a society where for centuries education was the traditional key to a better life. Dunfermline High School was formative: fiercely competitive, teachers really taught us how to think and all in a broad education. Even science stream pupils did french, greek and latin. Founded in 1468 it upheld its name for excellence. Being a "lad o'pairts" (an all-rounder) was admired in the Scotland of that day. It took some years to recognize these gifts I had been given.

Andrew Carnegie came from Dunfermline. Over the centuries, there was a tradition of people in Scotland striking out on their own and emigrating to other countries. And so it was for me.

Coming to America- Cameron: Cameron Iron Works recruited me to come to Houston from Scotland. I had done graduate work at University of Virginia and returned to UK for two years as required of Fulbright scholars. Going to America had been a dream while growing up. It led to a career in the petroleum industry. Now, as an American citizen and longtime Houston resident I am indebted to Cameron for giving me that start.

Herb Allen: Months later, while working at a drawing board in the engineering office at Cameron on Silber Road in Houston, a heavyset guy regularly visited me to see the latest on the design for a 200,000-ton forge press frame I was assigned to work on. He asked questions and made suggestions, really keeping me on my toes. He was an amazing engineer with a grasp of everything. People later told me who he was: Herb

Allen, long time president of the company. Though he had many other bigger things to worry about, he chose to dig into critical details. I also came to greatly admire his low-key positive approach to junior employees.

Unfortunately, a year later the team I had been in was disbanded. Most left, and I felt I had to find something else to do. After working with TOC, I started ETA. Years later, I learned of Herb Allen's major accomplishments in the oilfield and his generous philanthropy in Houston.

Jackup authority: In 1968-1969, I had an "apprenticeship" on jackup design at The Offshore Company (TOC). There I worked in the engineering department at their Houston office on Richmond Avenue. I reported to TOC's chief engineer, Tim Pease. There was a range of stimulating assignments. We did calculations on the *Constellation* class and the company's other jackups. There were many really talented engineers, a stimulating place to work with real engineering. What I learned at TOC helped shape what we did in ETA.

Computer analyses and cast steel joints: In the 1960s, pioneering structural analyses by computer for practical design was a new process that I learned from my first real job in UK in 1965-1966. I came to realize years later how formative that assignment was and how lucky I was to have that opportunity. Combined with the experience devising a steel structural connection system (which turned to be best as steel castings), it became a starting point for ETA's cast steel nodes for jackup legs.

No business training: In the UK, I worked for Tubewrights Ltd., and Stanley Rice, the managing director, was something of a mentor. He once said I really needed to get into business school. After coming to Houston, I filled in all the application forms and started the process for Harvard but came to the conclusion that I had had enough of schooling and could not stomach another two years in grad school. It was a mistake; ETA unintentionally became my business school!

Git-after-it: I learned that call for action early in arriving in Houston. In the late 2000s, I worked for Devon Energy, a large independent oil company in an offshore field development team alongside economists, petroleum engineers, geophysicists and drillers. The drillers were still *hands* like forty years ago and still said "git 'er done" when it came to taking action offshore. It was the old Texas version of Nike's "Just do it." Or, as American General George Smith Patton Jr. (1885-1945) said, "Lead, follow, or get out of the way." That was ETA.

Doing some good: A former colleague at ETA told me last year, "It was a fun place to work. People came in to the office on Saturday mornings." Over the years many ETA alumni have said similar things. They valued all they were able to do and learn and regretted ETA did not continue. The people were the big thing; they were talented and motivated people.

Lessons: There was a real marketing and management process at work in ETA, and I learned something about leadership and the business world. I admit I'm proud I was the force behind these ETA jackup designs and the other ideas that did not get built, inspiring that team of engineers. The ETA adventure was unlike anything I did before or after!

"Plus ca change, plus la meme chose": In 1965, *Drilling* magazine quoted the chairman of the board of Zapata Offshore Company. He discussed advances in the offshore rig fleet and closed with his industry outlook. His comments could in essence have been said today in 2018:

> *"In 1956, there were six barges capable of drilling in 35 ft. of water or less; three for 35-80 ft. Now there are seventeen for 35 ft. or less, twelve for 35-80 ft., and twenty eight for over 100 ft."*

There was big growth in the offshore fleet and its capabilities!

It was not without risks and losses; the article pictures the chairman smiling on receipt of what was then the biggest settlement check for the loss of a rig. It was $5.7 million for the drilling barge *Maverick*, which was lost during Hurricane Betsy. Who exactly was that chairman? He was George H.W. Bush, who became the forty-first president of the United States.

He closed on a thoughtful note:

> *When we look at our business as a whole, when we contemplate the challenges of the future and the ever-expanding worldwide demand, the problems don't seem so bad. I think we offshore contractors can cautiously say we've got our worries, but if the demand develops as predicted, "What's the problem?"*

Now in 2018, as optimism returns, what has really changed? Generations later, Bush's words are coming true. As the fleet and its capabilities have grown tremendously despite problems, what has really changed?

GLOSSARY

Abbreviation	Explanation
ABS	American Bureau of Shipping, the U.S. classifications society
AISC	American Institute for Steel Construction. Structural engineers widely used their handbooks and data in the 1970s.
BMC	Baker Marine Corporation with fabrication yard in Ingleside, Texas and jacks and leg chord plant at Cuddihy. Took over the U.S. assets of IHC Holland-R.G. LeTourneau around 1975, no longer building jackups.
BME	Baker Marine Engineers. An entity used by BMC to take over ETA's design assets and business after its collapse in 1977.
CDC	Control Data Corporation, the designer and builder of the CDC 6400 and 6600 supercomputers used in the 1970s by companies such as ETA.
CFEM	Compagnie Française d'Entreprises Métalliques in Dunkerque, France. Built the two ETA Europe Class jackups, went out of business around 1990.
COFACE	Compagnie Française d'Assurance pour le Commerce Extérieur) a financing plan to increase exports from France, the export credit agency for France since 1946.
COSL	China Oilfield Services Ltd., a majority controlled subsidiary of CNOOC (China National Offshore Oil Company).
DEEPWATER JACKUP	The first jackup design ETA developed. Created in 1971-1972, it had three legs with a slant leg configuration, able to work in up to 500 ft. of water. Too big and too early, it never got anywhere!
DESIGN SOURCE	A term that can be used for an independent design firm like ETA or an offshore drilling contractor with its own designs (like TOC) or an established builder like Bethlehem, Levingston, or LeTourneau.

As jackups were built in more parts of the world by more builders, the design source becomes a useful way of tracking the growth in use of different designs.

DNV — Det Norske Veritas, the Norwegian classification society, now known as DNV-GL.

DYVI — Jan Erik Dyvi was the chairman and founder of Dyvi Drilling A/S, a drilling contractor based in Oslo, Norway. Dyvi Drilling A/S retained ETA to provide its ETA Europe Class jackup design for construction of the *Dyvi Beta* and *Dyvi Gamma* jackups, delivered in 1976 and 1977.

EAGER BEAVER — A jackup design developed by ETA in 1975-1976 with three tubular legs and a mat for up to 180 ft. water depth. Though none were built, the design was a similar and competing concept to the Bethlehem units' three-legged mat jackups.

ETA — Engineering Technology Analysts, Inc., a Texas corporation formed in early 1970. Later rebranded as ETA Engineers, Inc.

FELS — Far East Levingston Shipbuilding Ltd. in Singapore. Was licensed in 1974 to build the ETA Asia Class jackup design. Over subsequent years, FELS became one of the world's leading builders of all kinds of MODUs.

HDW — Howaldtswerke Deutsche Werft, a majors shipbuilder in Kiel, Germany that delivered three jackups with six tubular legs during 1965-1976.

KIP — One kip is one kilo pound, or 1,000 lb., or one half of a short ton that is 2,000 lb.

KSI — A pressure or stress of one kip per square inch.

MOBILE MONOPOD — A jackup design concept for offshore drilling and production which was developed by ETA with a single central leg. Inspired four major designers to emulate it in the following four decades.

MODU — Mobile offshore drilling unit, e.g. a drill barge drillship, semisubmersible, submersible, or jackup

NKK	Nippon Kokan K.K., a major Japanese shipbuilder. NKK was interested in entering the offshore market in the mid-1970s, partnered with ETA in 1977, and transferred that connection to BME.
ONGC	Oil & Natural Gas Corporation, the national oil company of India.
OTC	Offshore Technology Conference, the world's leading conference for offshore, which has been held annually in Houston since 1969.
PE	Professional engineer, a license by the state to practice engineering.
POOL COMPANY	An onshore drilling and well-servicing company in San Angelo, Texas. It entered the offshore market in the early 1970s and retained ETA to provide jackup design services for their 140 and then 50 series of jackups. Later became part of Nabors.
RACK CHOCKS	A device to transfer vertical loads from the jackup leg directly into the hull (rather than relying on the holding power of the elevating system). Developed in the late 1970s.
RDL	Redpath Dorman Long, an offshore fabricator in Methil, Fife, Scotland that proposed fabrication of ETA Europe Class jackups in 1974. Still in business today.
RFP	Request for Proposal.
ROBRAY	Robray Offshore Drilling Co. Ltd., based in Singapore. ETA's first jackup design client in 1973 with the ETA Robray 300 Class design.
SCANDRIL	A Houston-based jackup design firm backed by Navire of Finland that offered jackup designs, competed with ETA during 1974-1977.
SNAME	Society of Naval Architects and Marine Engineers. Active in ETA's time and today.
SPE	Society of Petroleum Engineers. Active in ETA's time and today.
SPUD TANK	The tank at the bottom of a jackup leg which penetrates into the sea floor.
SWRI	Southwest Research Institute, located in San Antonio, Texas. An R&D organization often serving the petroleum industry, it investigated stress concentration factors on offshore structures in

	the 1970s and 1980s.
TAPS	Trans-Alaska Pipeline System.
TEXACO	A major oil company now integrated into Chevron.
TEXAS A&M	Texas A&M University, home of the Aggies. A source of excellent engineering talent, though it was often the butt of Aggie jokes in the 1970s.
TOC	The Offshore Company, later known as Sonat and now as Transocean.
TON	Can be either short or long: a "short" ton is 2,000 lb., a unit of weight commonly used in the U.S. A "long" ton is 2,240 lb. often used in marine circles as a unit of displacement. The long ton originated in U.K.
TONNE	A metric ton = 1,000 kilograms = 2,202 lb.
UNIVAC	Designer and builder of supercomputers in the 1970s. The Univac 1108 was often used by ETA in its analyses.
ZENTECH	A Houston-based company founded by two engineers from ETA. It has grown to be a leading jackup engineering authority and designer.

ACKNOWLEDGEMENTS

As mentioned in the introduction, the idea for a story on the ETA jackups originated with Tim Pease in 2005. In January 2017, when he and Dr. Malcolm Sharples combined and started asking penetrating questions, I began assembling a narrative on the engineering principles used in ETA's pioneering jackup designs, and it grew and grew to be book-length.

It has often been difficult to track down historical details on the series of jackups because the events discussed here occurred a long time ago, and many people who were involved have left the industry, retired, or passed away.

I am indebted to friends in the industry for their recollections on facts and the underlying thoughts behind past events. I thank them for their frank debate, inputs, pictures, and help in making this book as accurate, fair, and balanced as possible.

Ian Craven and Richard Greff provided the necessary perspective about Robray Offshore Drilling Co. from 1970s and onwards. They based this perspective on their past employment there, and their insight was appreciated, as few in the industry knew much about this Singapore-based company.

I am grateful to Kjell Evensgard of Maersk Drilling, who checked my comparison data in Table S and provided the images of *Maersk Intrepid* in Picture 27 for what is now the world's biggest jackup.

Dr. Rao Guntur, Ramesh Maini, Ralph McTaggart, and Derek Scovell were colleagues at ETA who are still in Houston. Over numerous conversations and lunches, they patiently helped me fill in and get the ETA story right. All four were recently-arrived immigrants decades ago; Dr. Rao Guntur (then a recent graduate of University of Texas) and Ramesh Maini were natives of India, Ralph McTaggart came from Scotland, and Derek Scovell was originally from England.

John Irwin read an early draft of this saga and emphasized the importance of telling this story. That was powerful encouragement coming from someone with his long career in the leadership of offshore drilling, which included his contribution to Atwood Oceanics, where he worked from 1992-2009 and rose to become president and chief executive officer. He was also an immigrant who came to America about the same time I did but from farther away— Australia. In our discussions I learned about his dedication to educate the next generation; he was the 2011 Brenton S. Halsey Distinguished Visiting Professor at the University of Virginia. I am greatly indebted to him for writing his thoughtful foreword.

I was fortunate to be put in touch with a historian of the rig building business in Singapore, Waiman Kwan, who provided several internet images and informed me on subjects I had not known about.

In ETA's days, Carl Wendenburg was a Houston-based engineer with ETA's biggest jackup client that later became a competitor (LeTourneau). He has had a stellar career in the drilling equipment business and gave wise feedback.

Table T, on the next page, lists these industry veterans, their roles in the ETA era of 1970-1975, and what they are doing today.

Great recognition and thanks is due to Shanley McCray and her team at Opportune Independent Publishing of Houston who patiently translated my manuscript into this book and succeeded in making all these pictures, diagrams and tables intelligible.

Name	Role during the ETA era of 1970-1975	Role today in 2018
Ian Craven	Materials Manager, Robray Offshore Drilling Co., Burma	Owner, Icarus Consulting, Cebu City, Philippines
Kjell Evensgard	None	Head of Sales, Maersk Drilling Norge
Richard Greff	Engineer, ETA, Houston / Engineer, Robray, Singapore	Retired, Texas
Dr. Rao Guntur	Engineering analyst, ETA, Houston	Founder, EVP, Zentech, Houston
John Irwin	Management engineer, Kerr McGee, Oklahoma City	Company Director, Advisor, ex CEO of Atwood Oceanics
Ramesh Maini	Engineering Analyst, ETA, Houston	Founder, Chairman, Zentech, Houston
Ralph McTaggart	Chief Naval Architect, ETA, Houdton	Retired, Houston
Tim Pease	Chief Engineer, The Offshore Company, Houston	Consultant, US Onshore Offshore, Houston
Dr. Malcolm Sharples	Consultant, Noble Denton, London	Consultant, Offshore Technology & Risk, Houston
Derek Scovell	Project Manager, ETA, Houston	Retired, Houston
Kwan Waiman	Shipyard staff, Singapore	Historian on Singapore rig building, Singapore
Carl Wendenburg	Engineer, Marathon LeTourneau, Houston	Managing Director, Rig Masters Pte. Ltd., Singapore

<u>Table T:</u>
Contributors to this history: Their roles in the ETA days and today

APPENDIX—RELEVANT PUBLICATIONS

This list covers difficult-to-find articles published by ETA in the 1970s, as well as copies of the five U.S. patents granted to ETA (as mentioned in the text). The first eighteen publications listed below are free to download at www.ETAanditsJackups.com

<u>By ETA's employees or about ETA</u> (chronological order)

1. Lovie, P.M.; Lowery, E.L.: "Jack-up for 400 ft. Water Depths Feasible," <u>Oil & Gas Journal</u>, 10 January 1972, pp. 86-87.

2. Lovie, P.M.: "Why Not a Jack up for 400 ft. Water?" Engineering Technology Analysts, Inc., <u>Innovation</u>, volume 1, number 1, first quarter 1972, 3 pages.

3. Lovie, P.M.: "Spud Tank for Offshore Drilling Unit," <u>United States Patent & Trademark Office</u>, no. 3,823,563, filed 5 September 1972, granted 16 July 1974, 7 pages.

4. Lovie, P.M.: "Self-Elevating Offshore Platform with Folding Legs," <u>United States Patent & Trademark Office</u>, no. 3,826,099, filed 25 September 1972, granted 30 July 1974, 7 pages.

5. Lovie, P.M.: "The Robray 300—The First of a New Generation of Jack-ups to be Built," Engineering Technology Analysts, Inc., <u>Innovation</u>, volume 2, number 2, 1973, 2 pages.

6. Lovie, P.M.; Lowery, E.L.: "Means for Altering Motion Response of Offshore Drilling Units," <u>United States Patent & Trademark Office</u>, no. 3,916,633, filed 24 August 1973, granted 4 November 1975, 5 pages.

7. Gunderson, R.H.; McTaggart, R.G.: "On Damaged Stability of Drilling Vessels," Society of Naval Architects & Marine Engineers, Annual Meeting, New York, November 15-17, 1973, 13 pages.

8. Lowery, E.L.: "Mobile Marine Drilling Unit," <u>United States Patent & Trademark Office</u> no. 3,996,754, filed December 14, 1973, granted December 14, 1976, 7 pages.

9. Angerstein, J.: "Jack-ups-a Future in the North Sea," interview, <u>Northern Offshore</u>,

April 1974, pp. 33-42.

10. "New Innovative Designs for Jackups for the North Sea Oil Patch," <u>Offshore</u>, April 1974, 4 pages.

11. Lovie, P.M.: "Self-Elevating Offshore Drilling Unit Legs," <u>United States Patent</u>, no. 3,967,457, filed 11 July 1974, granted 6 July 1976, 8 pages.

12. Bynum, D.; Lovie, P.M.: "How Jack-ups Fit in the North Sea Boom," <u>Petroleum Engineer</u>, October 1974, 6 pages.

13. Bynum, D.; Lovie, P.M.: "A Thousand Rigs Could be Needed to Meet World Goals," <u>Offshore</u>, January 1975, pp. 53-60. [Title chosen by magazine!!!]

14. Bynum, D.; Lovie, P.M.: "Rig Safety is Vital," <u>Offshore</u>, May 1975, 8 pages.

15. Lovie, P.M.: "MODU Classification and Certification," <u>The Technology of Offshore Drilling, Completion and Production</u>, Pennwell Publishing, 1976, 426 p., pp. 389-413.

Pioneering discussion of the jackup kit concept of design+jacks+leg chords

16. Lovie, P.M.: "Jack-up Kits for Offshore Plants to be Constructed Anywhere in the World," Symposium on Coastal & Offshore Plants, Washington DC, Coastal & Offshore Plant Systems, 25-26 January 1977, 17 pages.

Precedent that inspired the cast steel nodes

17. Lovie, P.M.; Rice, S.F; Scholfield, K.; Somerset, F.K.: " Method of Creating Metal Structures and jointing devices for use therein," <u>Tubewrights Limited</u>, British patent application no. 26861/64, filed 29 June 1964, completed 28 June 1965, 24 pages, 16 Figures.

18. Lovie, P.M.: "Peter's First Real Job," 6 pages, August 2016, available on line at http://www.lovie.org/pdf/peters-first-real-job.pdf

[This is a collection of materials from 1965-1966 on the structural analysis, design, and full-scale testing of a prototype: a 152-ft.-high geodetic electricity transmission tower that used tubular structural members and cast steel connection pieces. This text includes a summary of the project description, digital recoveries of faded 1965 photographs, excerpts from the worldwide patent application, and the international conference paper that reported on the project.]

<u>Third-party information sources</u>

19. Rechtin, E.C.; Scales, R.E.; Steele, J.E.: "Engineering Problems Related to the Design of Offshore Mobile Platforms," <u>Society of Naval Architects and Marine Engineers</u> (SNAME), Annual Meeting 14-17 November 1957, 49 pages.

20. Bush, G.H.W.: "The Future of the Offshore Drilling Industry," <u>Drilling</u>, November 1965, 3 pages, based on the October 1965 paper presentation to then Houston chapter of API.

21. "Rules for Building and Classing Offshore Mobile Drilling Units," <u>American Bureau of Shipping</u>, 1973, 152 pages.

22. Det Norske Veritas: "Rules for the Construction and Classification of Mobile Offshore Drilling Units, 1975.

23. Armstrong, J.E.: "Leg Structure for Jack-up Platform with Single Point Jacking," <u>United States Patent & Trademark Office</u>, no. 4,160,538, filed 2 May 1977, awarded 10 July 1979, 6 pages.

24. Laplante, G.; Clauw, R.: "Trigone Jack-up Drilling Rig," <u>Offshore Technology Conference</u>, paper 3244, 1978, 8 pages.

25. Tsutomunu Nakajima et al, "Jackup Rig Leg Node Fatigue Testing," <u>Oil & Gas Digest</u>, pp. 10-11, September 1980.

26. Herrmann, R.P.; Pease, F.T.; Ray, D.R.: "Gravity Base, Jack-up Platform – Method and Apparatus," <u>United States Patent & Trademark Office</u>, no. 4,265,568, filed 6 August 1979, awarded 5 May 1981, 18 pages.

27. Cojeen, H.P.; Johnson, R.E.: "An Investigation into the Loss of the Mobile Offshore

Drilling Unit Ocean Ranger," Marine Technology, Vol. 22. No. 2, April 1985, pp. 109-125.

28. Steele, J.E.: "Mobile Offshore, Jack-up, Marine Platform Adjustable for Sloping Sea Floor," United States Patent & Trademark Office no. 4,668,127, filed 22 April 1986, awarded 26 May 1987, 19 Pages.

29. Smith, D.J.: "Project Management of Subsidence and Ekofisk Elevating Project," Offshore Technology Conference, paper no. 5655, 1988, 18 pages.

30. Perol, C.; Rodgers, S.: "The Jackup Monopile: A Production Drilling Platform," Offshore Technology Conference, paper no. 6614, 1991, 13 pages.

31. Arnesen, C.A.; Kjeey, H.; Eriksson, H.: "Structural Behavior of Harsh Environment Jack-ups," "The Jackup Drilling Platform," Gulf Publishing, chapter 7, pp. 90-136, 1985.

32. Hooper, R.B.: "The Bush Family and the Offshore Oil Industry," Go Gulf magazine, March/April 2000, pp. 38-41.

33. "Guidelines for Site Specific Assessment of Mobile Jackups." Society of Naval Architects & Marine Engineers, Technical & Research Bulletin 5-5A, Revision 3, August 2008, 366 pages.

34. "Petroleum and natural gas industries — Site-specific assessment of mobile offshore units—Part 1: Jack-ups," International Standard ISO 19905-1, First edition, 2012-08-01, 381 pages. International Organization for Standardization.

35. Bauer, R.F.; Schempf, F.J.: "Roughnecking: My Life & Times as a Cowboy, Petroleum Engineer and Offshore Pioneer," Surge Press, 180 pages, 2009.

36. William L. Leffler, W.L.: Pattarozzi, R.; Sterling, G.: " Deepwater Petroleum Exploration & Production: A Nontechnical Guide," Pennwell Publishing, 350 pages, 2011.

37. "The International Conference: The Jackup Platform, School of Engineering and Mathematical Sciences," City University of London, 19-20 September 2017, held biennially.

www.ingramcontent.com/pod-product-compliance
Lightning Source LLC
Chambersburg PA
CBHW082009190326
41458CB00010B/3137